T0191497

Methods in Molecular Biology

Series Editor
John M. Walker
School of Life and Medical Sciences
University of Hertfordshire
Hatfield, Hertfordshire, AL10 9AB, UK

For further volumes:
http://www.springer.com/series/7651

The Nuclear Receptor Superfamily

Methods and Protocols

Second Edition

Edited by

Iain J. McEwan

School of Medicine, Medical Sciences and Nutrition, Institute of Medical Sciences, University of Aberdeen, Aberdeen, Scotland, UK

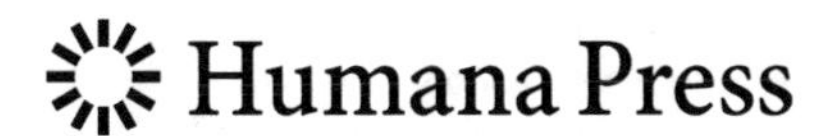

Editor
Iain J. McEwan
School of Medicine
Medical Sciences and Nutrition
Institute of Medical Sciences
University of Aberdeen
Aberdeen, Scotland, UK

ISSN 1064-3745 ISSN 1940-6029 (electronic)
Methods in Molecular Biology
ISBN 978-1-4939-8120-5 ISBN 978-1-4939-3724-0 (eBook)
DOI 10.1007/978-1-4939-3724-0

Softcover reprint of the hardcover 2nd edition 2016

Printed on acid-free paper

This Humana Press imprint is published by Springer Nature
The registered company is Springer Science+Business Media LLC New York

Preface

It is now 30 years since the first steroid receptor cDNAs were cloned, a development that led to the concept of a superfamily of ligand-activated transcription factors: the nuclear receptors. Nuclear receptors share a common architecture at the protein level, but a remarkable diversity is observed in terms of natural ligands and xenobiotics that bind to and regulate receptor function. Natural ligands for nuclear receptors are generally lipophilic in nature and include steroid hormones, bile acids, fatty acids, thyroid hormones, certain vitamins, and prostaglandins. A significant proportion of the family members have been described as *orphans*, as the natural ligand, if it exists, remains to be identified. Nuclear receptors act principally to directly control patterns of gene expression and play vital roles during development and in the regulation of metabolic and reproductive functions in the adult organism. Since the original cloning experiments, considerable progress has been made in our understanding of the structure, mechanisms of action, and role in disease of this important family of proteins. This volume of *Methods in Molecular Biology* follows on from an earlier edition (Volume 505) and aims to describe a complementary range of molecular, cell biological, and in vivo protocols used to investigate the structure–function of nuclear receptors, together with experimental approaches that may lead to new drugs to selectively target nuclear receptor-associated diseases. This volume will be of great benefit and use to those starting out in the nuclear receptor research field (life sciences graduate students and postdoctoral fellows) as well as to more established researchers who wish to apply different methods to a particular receptor/research problem. The volume will also be of use to medical students and clinicians undertaking research in this ever-growing field of study.

Aberdeen, UK *Iain J. McEwan*

Contents

PART IV MODEL SYSTEMS

Contributors

STEFAN J. BARFELD • *Prostate Cancer Research Group, Centre for Molecular Medicine Norway (NCMM), Nordic EMBL Partnership University of Oslo and Oslo University Hospital, Oslo, Norway*
SILVÈRE BARON • *Université Clermont Auvergne, Université Blaise Pascal, Génétique Reproduction et Développement, Clermont-Ferrand, France; CNRS, UMR 6293, GReD, Aubière, France; INSERM, UMR 1103, GReD, Aubière, France; Centre de Recherche en Nutrition Humaine d'Auvergne, Clermont-Ferrand, France*
SOUMYA BASU • *Molecular Functional Imaging Lab, ACTREC, Tata Memorial Centre, Navi Mumbai, India*
BAGORA BAYALA • *Université Clermont Auvergne, Université Blaise Pascal, Génétique Reproduction et Développement, Clermont-Ferrand, France; CNRS, UMR 6293, GReD, Aubière, France; INSERM, UMR 1103, GReD, Aubière, France; Centre de Recherche en Nutrition Humaine d'Auvergne, Clermont-Ferrand, France; University of Koudougou, Burkina, Faso*
CLAUDE BEAUDOIN • *Université Clermont Auvergne, Université Blaise Pascal, Génétique Reproduction et Développement, Clermont-Ferrand, France; CNRS, UMR 6293, GReD, Aubière, France; INSERM, UMR 1103, GReD, Aubière, France; Centre de Recherche en Nutrition Humaine d'Auvergne, Clermont-Ferrand, France*
CHARLOTTE L. BEVAN • *Androgen Signalling Laboratory, Department of Surgery and Cancer, Imperial College London, London, UK*
ARTEM CHERKASOV • *Vancouver Prostate Centre, University of British Columbia, Vancouver, BC, Canada*
ANNE T. COLLINS • *YCR Cancer Research Unit, Department of Biology, University of York, Heslington, North Yorkshire, UK*
KUSH DALAL • *Vancouver Prostate Centre, University of British Columbia, Vancouver, BC, Canada*
D. ALWYN DART • *Androgen Signalling Laboratory, Department of Surgery & Cancer, Imperial College London, London, UK; The Cardiff China Medical Research Collaborative, Cardiff University School of Medicine, Cardiff, UK*
ABHIJIT DE • *Molecular Functional Imaging Lab, ACTREC, Tata Memorial Centre, Navi Mumbai, India*
SCOTT M. DEHM • *Masonic Cancer Center and Department of Laboratory Medicine and Pathology, University of Minnesota, Twin Cities, MN, USA*
KARA A. DESANTIS • *State University of New York at New Paltz, New Paltz, NY, USA*
SHALINI DIMRI • *Molecular Functional Imaging Lab, ACTREC, Tata Memorial Centre, Navi Mumbai, India*
FIONA M. FRAME • *YCR Cancer Research Unit, Department of Biology, University of York, Heslington, North Yorkshire, UK*
KATI KIVINUMMI • *Prostate Cancer Research Center, Institute of Biosciences and Medical Technology – BioMediTech and Fimlab Laboratories, University of Tampere and Tampere University Hospital, Tampere, Finland*

Nada Lallous • *Vancouver Prostate Centre, University of British Columbia, Vancouver, BC, Canada*
Leena Latonen • *Prostate Cancer Research Center, Institute of Biosciences and Medical Technology – BioMediTech and Fimlab Laboratories, University of Tampere and Tampere University Hospital, Tampere, Finland*
Yingming Li • *Masonic Cancer Center and Department of Laboratory Medicine and Pathology, University of Minnesota, Twin Cities, MN, USA*
Jean-Marc A. Lobaccaro • *Université Clermont Auvergne, Université Blaise Pascal, Génétique Reproduction et Développement, Clermont-Ferrand, France; CNRS, UMR 6293, GReD, Aubière, France; INSERM, UMR 1103, GReD, Aubière, France; Centre de Recherche en Nutrition Humaine d'Auvergne, Clermont-Ferrand, France*
Norman J. Maitland • *YCR Cancer Research Unit, Department of Biology, University of York, Heslington, North Yorkshire, UK*
Charles E. Massie • *Li Ka Shing Centre, Cancer Research UK Cambridge Institute, University of Cambridge, Cambridge, UK*
Iain J. McEwan • *School of Medicine, Medical Sciences and Nutrition, Institute of Medical Sciences, University of Aberdeen, Aberdeen, Scotland, UK*
Dagmara McGuinness • *Wolfson Wohl Translational Research Centre, Institute of Cancer Sciences, University of Glasgow, Glasgow, UK*
Ian G. Mills • *Prostate Cancer Research Group, Centre for Molecular Medicine Norway (NCMM), Nordic EMBL Partnership University of Oslo and Oslo University Hospital, Oslo, Norway; Department of Molecular Oncology, Institute of Cancer Research and Oslo University Hospital, Oslo, Norway; Prostate Cancer Research Group, Centre for Cancer Research and Cell Biology (CCRCB), Queen's University of Belfast, 97 Lisburn Road, Belfast BT9 7AE, UK*
Hélène Morin • *Vancouver Prostate Centre, University of British Columbia, Vancouver, BC, Canada*
Ravi Munuganti • *Vancouver Prostate Centre, University of British Columbia, Vancouver, BC, Canada*
Laura O'Hara • *MRC Centre for Reproductive Health, University of Edinburgh, Edinburgh, UK*
Davide Pellacani • *YCR Cancer Research Unit, Department of Biology, University of York, Heslington, North Yorkshire, UK*
Jeffrey L. Reinking • *State University of New York at New Paltz, New Paltz, NY, USA*
Paul S. Rennie • *Vancouver Prostate Centre, University of British Columbia, Vancouver, BC, Canada*
Mauro Scaravilli • *Prostate Cancer Research Center, Institute of Biosciences and Medical Technology – BioMediTech and Fimlab Laboratories, University of Tampere and Tampere University Hospital, Tampere, Finland*
Fred Schaufele • *Center for Reproductive Sciences, University of California San Francisco, San Francisco, CA, USA*
Lee B. Smith • *MRC Centre for Reproductive Health, University of Edinburgh, Edinburgh, UK*
Amalia Trousson • *Université Clermont Auvergne, Université Blaise Pascal, Génétique Reproduction et Développement, Clermont-Ferrand, France; CNRS, UMR 6293, GReD, Aubière, France; INSERM, UMR 1103, GReD, Aubière, France; Centre de Recherche en Nutrition Humaine d'Auvergne, Clermont-Ferrand, France*
Tapio Visakorpi • *Prostate Cancer Research Center, Institute of Biosciences and Medical Technology – BioMediTech and Fimlab Laboratories, University of Tampere and Tampere University Hospital, Tampere, Finland*

Part I

Introduction

Chapter 1

The Nuclear Receptor Superfamily at Thirty

Iain J. McEwan

Everything I know I learned after I was thirty

George Clemenceau

The body is at its best between the ages of thirty and thirty-five

Aristotle

Abstract

The human genome codes for 48 members of the nuclear receptor superfamily, half of which have known ligands. Natural ligands for nuclear receptors are generally lipophilic in nature and include steroid hormones, bile acids, fatty acids, thyroid hormones, certain vitamins, and prostaglandins. Nuclear receptors regulate gene expression programs controlling development, differentiation, metabolic homeostasis and reproduction, in both a temporal and a tissue-selective manner. Since the original cloning of the cDNAs for the estrogen and glucocorticoid receptors, large strides have been made in our understanding of the structure and function of this family of transcription factors and their role in pathophysiology.

Key words Steroid hormones, Nuclear receptors, Gene expression, Allosteric regulation

---

The cloning of the cDNAs for nuclear receptors (NRs) [1] opened up a whole new chapter in research into the regulation of cell function and metabolism. In the human genome there are 48 known nuclear receptor genes, which participate in a wide range of physiological processes, from the control of reproduction to the regulation of metabolism and development [2]. Half of these receptor proteins are known to bind an identifiable ligand (Fig. 1a). As a consequence of the importance of nuclear receptors for both health and disease they have long been recognized as validated drug targets in pathophysiology conditions including hormone-dependent cancers, inflammation and metabolic syndrome. A major recent goal has been the identification

Iain J. McEwan (ed.), *The Nuclear Receptor Superfamily: Methods and Protocols*, Methods in Molecular Biology, vol. 1443,
DOI 10.1007/978-1-4939-3724-0_1,

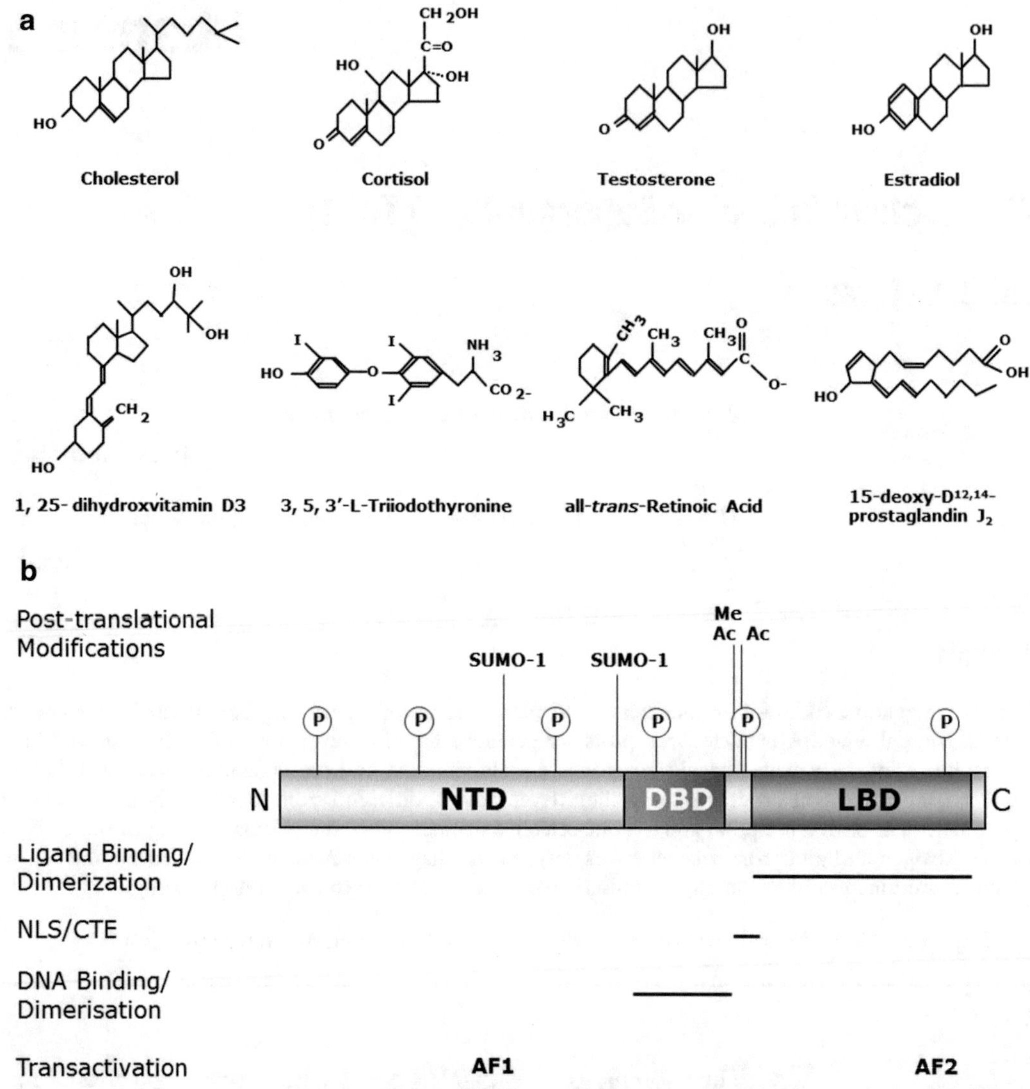

Fig. 1 Of ligands and receptors. (**a**) A selection of known ligands for members of the nuclear receptor superfamily. These include cholesterol and its metabolites, steroid hormones (cortisol, testosterone, estradiol) and vitamin D_3 (*top row*). Other ligands are derived from amino acids (thyroid hormone) and lipid and fatty acid metabolites (retinoic acid, prostaglandin J_2). (**b**) Schematic showing the functional and structural domain organization of nuclear receptors: *LBD* ligand-binding domain, *DBD* DNA-binding domain, and *NTD* amino terminal domain. Also illustrated are activation functions (AF) 1 and 2, sequences important for nuclear localization (NLS) and DNA binding (C-terminal extension *CTE*). Above the domain organization are types of posttranslational modification most commonly associated with nuclear receptors

and development of selective agonists or antagonists with improved therapeutic properties and reduced side effects.

1 Nuclear Receptor Domain Organization and Function

1.1 DNA-Binding Domain

NRs nearly all share the same canonical protein structure (Fig. 1b) [2, 3]. The DNA-binding domain (DBD) conforms to a Zn-finger subtype of transcription factors, where eight highly conserved cysteine residues coordinate two zinc ions and also mediate homo- or hetero-dimerization (Fig. 1b). The receptors for androgens (AR), corticosteroids (GR, MR), and progesterone (PR) or estrogens (ER) typically bind DNA sequences configured as inverted repeats of the half-site sequences AGGACA or AGGTCA, respectively, while the remaining members of the family bind half-sites, of the sequence AGGTCA, arranged singly or as direct repeats [3, 4]. The advent of next-generation sequencing, in combination with chromatin immunoprecipitation (ChIP-seq), has allowed for the identification of NR-binding sites in a genome-wide manner and in different cell types and under different pathophysiological conditions (reviewed in [5]). Furthermore these studies have revealed that NR response elements can vary from monomeric half-sites to nearly perfect inverted or direct repeats, and have demonstrated the cooperation between NRs and other transcription factors, including FOXA1 (AR, ER), AP-1 (GR, liver X receptor (LXR)), and C/EBP (peroxisome proliferator-activated receptor (PPAR)-γ, LXR) ([5] and references therein). ChIP methods have also been used to look at genome-wide changes in histone markers of gene repression (methylation) and activation (acetylation, methylation) in response to NR signaling. In further developments the transcriptional response of the AR and GR, as a consequence of SUMOylation, has been investigated by ChIP-seq, revealing an unexpected selectivity in the regulation of target genes [6, 7].

1.2 Ligand-Binding Domain

The C-terminal domain of NRs binds agonists or antagonists, with the ligand-binding pocket buried within a highly helical globular domain. The ligand-binding pocket (LBP) ranges in volume from 0 to 1500 $Å^3$ and is able to accommodate cholesterol, steroid hormones, or products of intracellular metabolism (Fig. 1a) [3, 8]. For those NRs where a ligand is thought to bind, ligand binding results in a structural rearrangement and the formation of a hydrophobic pocket on the surface (termed AF2) responsible for co-regulatory protein binding. In addition, folding of the globular LBD creates a surface termed BF3 [9]. This surface pocket has generated considerable interest due to the potential for the binding of small-molecule modulators and the allosteric regulation between the LBP, AF2 and BF3 (reviewed in Ref. 9). Surfaces of the LBDs can also participate in dimerization between receptor monomers (Figs. 1b and 2).

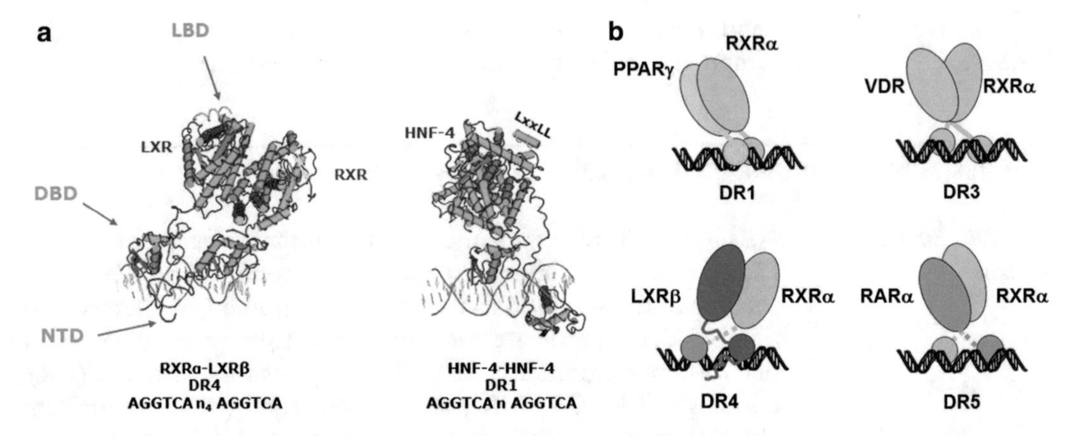

Fig. 2 Nuclear receptor complexes. (**a**) Crystal structures of the RXRα-LXRβ heterodimer (4NQA) and the homodimer of HNF-α (4IQR) bound to DR4 [16] and DR1 [19] DNA response elements, respectively. (**b**) Schematic representations of RXR heterodimer complexes on DR1, 3, 4, and 5 DNA elements (*see* text for details)

1.3 Amino-Terminal Domain

The N-terminal domain (NTD) is the least well-conserved domain and is the largest for the steroid receptor subfamily. The NTD contains sequences necessary for transcriptional regulation (termed AF1), which can operate independently of ligand when linked only to a DNA-binding domain (Fig. 1b). This region has been characterized as intrinsically disordered, with folding coupled to protein-protein interactions and receptor function (reviewed in Refs. 10 and 11). Intrinsic disorder can be thought of as an ensemble of conformers representing structural plasticity and which range from fully unfolded to varying levels of secondary (α-helices, β-strands) and tertiary folding [10].

1.4 Hinge Region

The AR, GR, and PR are generally described as undergoing a cytoplasm to nuclear translocation upon hormone binding. The other members of the NR family are thought to be primarily nuclear and in the majority of cases may already be bound to DNA response elements (e.g., RAR, TR). A bipartite nuclear localization sequence has been mapped to the hinge region (Fig. 1b). The hinge region also plays an important role in inter-domain communications and allosteric regulation (reviewed in Ref. 10) (see also below).

2 Regulation of Receptor Activity by Posttranslational Modifications

In addition to regulation of NR action by ligand, receptors are also subject to a plethora of posttranslational modifications, including phosphorylation, sumolyation, acetylation, and methylation (Fig. 1b) (reviewed in Refs. 12–14). The addition of different chemical groups can regulate receptor protein stability, intracellular location, and DNA-binding properties and allows for the cross talk between NRs and cell surface receptor signaling pathways.

3 Structural Analysis of Nuclear Receptor Complexes

The cloning of NRs not only opened up the possibility to better understand receptor biology and roles in physiology, through knockout and knock-down approaches, but has also allowed detailed molecular and structural analysis. In the last 5 years we have seen the advance from the structures of isolated domains (LBD, DBD) to full-length or nearly full-length receptor complexes (Fig. 2) (reviewed in Ref. 15). These studies have confirmed the globular structures of the LBD and DBD and emphasized the structural flexibility of the NTD, as this region is often deleted on absent from the final structural model. A notable exception is the structure of RXRα-LXRβ heterodimer bound to DNA, where a portion of the LXR–NTD was observed interacting with the 5′-half-site of the response element (Fig. 2a) [16]. Other common features of the complexes solved so far are the "open" conformation of the receptor complexes on DNA and the role of the hinge region (between the LBD and DBD) in intra-domain interactions (Fig. 2b). Solution structures, using small-angle X-ray scattering (SAX), FRET, and hydrogen-deuterium exchange (HDX), as well as the X-ray structure of the LXR complex, revealed this extended conformation for complexes involving the heterodimer partner RXR, which occupies the 5′ half-site on DR3, 4, and 5 elements and the 3′ half-site on a DR1 DNA response element (Fig. 2b) [16–18]. The only high-resolution structure for a homo-dimer complex so far reported is for hepatocyte nuclear factor (HNF)-4 (Fig. 2a) [19], although a low-resolution cryo-EM structure of the ERα has recently been published [20]. The HNF-4 structure is particularly noteworthy as it provides a structural basis for disease-causing mutations and also highlights how posttranslational modifications (serine-phosphorylation and arginine-methylation) at sites not directly contacting DNA can regulate receptor DNA binding and function [19].

In this volume of Methods in Molecular Biology, a companion to volume 505, a number of methods are described for investigating NR action, both at the molecular and whole organismal levels. In addition the reader is directed to a number of Web-based resources for nuclear receptor research (*see* **Notes 1–4**).

In Part I there are three chapters illustrating methods for identifying NR ligands (Lombaccaro and Reinking and co-workers) and a protocol for drug discovery of AR antagonists (Cherkasov and co-workers).

In Part II, the protocols focus on integrating NR mechanisms of action. De and co-workers and Schaufele described methods for "imaging" receptor action and protein-protein interactions in living cells. In Chapter 7 McGuiness and McEwan describe the use of phospho-specific antibodies to determine site-specific phosphorylation of the androgen receptor in cell models and tissue samples.

The ability to study the genome-wide action of NRs has dramatically increased our understanding and knowledge of receptor-dependent gene regulation, from the identification of target genes to descriptions of different DNA response element architectures. Chapters 8 (Barfeld and Mills) and 9 (Massie) describe recent adaptations of chromatin immunoprecipitation (ChIP) methods for investigating chromatin structure and protein–DNA interactions.

In Chapter 10, Latonen and co-workers describe the identification of microRNAs regulated by the AR. miRNAs are now known to both regulate the expression of genes for NRs and mediate the action of NRs at target gene loci. Similarly, great attention is now being paid to splice variants and isoforms of NRs which may modulate normal receptor function or drive disease processes. Li and Dehm (Chapter 11) describe protocols for identifying and quantifying variants of the AR that lack the LBD.

In Part III, the focus turns to in vivo models for investigating NR function. Maitland and co-workers (Chapter 12) describe methods for isolating and culturing primary cells from patient tumor samples. Dart and Bevan (Chapter 13) look at visualizing receptor activity using a luciferase reporter gene and whole-animal imaging. Lastly, O'Hara and Smith (Chapter 14) describe the power of selective targeting of NRs in tissue-specific manner to help elucidate the cell-specific actions of, in this case, the steroid receptor for androgens.

4 Notes

1. https://www.nursa.org/nursa/index.jsf
2. http://www.receptors.org/nucleardb/
3. http://endocrinedisruptome.ki.si/
4. http://nrdbs.ucr.edu/binplone

References

1. Evans RM (1988) The steroid and thyroid hormone receptor superfamily. Science 240: 889–895
2. Germain P, Staels B, Dacquet C, Spedding M, Laudet V (2006) Overview of nomenclature of nuclear receptors. Pharmacol Rev 58: 685–704
3. Rastinejad F, Huang P, Chandra V, Khorasanizadeh S (2013) Understanding nuclear receptor form and function using structural biology. J Mol Endocrinol 51:T1–T21
4. Helsen C, Claessens F (2014) Looking at nuclear receptors from a new angle. Mol Cell Endocrinol 382:97–106
5. Meyer CA, Tang Q, Liu XS (2012) Minireview: applications of next-generation sequencing on studies of nuclear receptor regulation and function. Mol Endocrinol 26:1651–1659
6. Paakinaho V, Kaikkonen S, Makkonen H, Benes V, Palvimo JJ (2014) SUMOylation regulates the chromatin occupancy and anti-proliferative gene programs of glucocorticoid receptor. Nucleic Acids Res 42:1575–1592
7. Sutinen P, Malinen M, Heikkinen S, Palvimo JJ (2014) SUMOylation modulates the transcriptional activity of androgen receptor in a target gene and pathway selective manner. Nucleic Acids Res 42:8310–8319

8. Gallastegui N, Mackinnon JA, Fletterick RJ, Estebanez-Perpina E (2015) Advances in our structural understanding of orphan nuclear receptors. Trends Biochem Sci 40:25–35
9. Buzon V, Carbo LR, Estruch SB, Fletterick RJ, Estebanez-Perpina E (2012) A conserved surface on the ligand binding domain of nuclear receptors for allosteric control. Mol Cell Endocrinol 348:394–402
10. Kumar R, McEwan IJ (2012) Allosteric modulators of steroid hormone receptors: structural dynamics and gene regulation. Endocr Rev 33:271–299
11. McEwan IJ (2012) Intrinsic disorder in the androgen receptor: identification, characterisation and drugability. Mol Biosyst 8:82–90
12. Gioeli D, Paschal BM (2012) Post-translational modification of the androgen receptor. Mol Cell Endocrinol 352:70–78
13. Anbalagan M, Huderson B, Murphy L, Rowan BG (2012) Post-translational modifications of nuclear receptors and human disease. Nucl Recept Signal 10:e001
14. Berrabah W, Aumercier P, Lefebvre P, Staels B (2011) Control of nuclear receptor activities in metabolism by post-translational modifications. FEBS Lett 585:1640–1650
15. McEwan IJ, Kumar R (2015) Twenty-five years of nuclear receptor structure analysis: from laboratory to clinic. In: McEwan IJ, Kumar R (eds) Nuclear receptors: from structure to clinic. Springer, New York, NY, pp 1–14
16. Lou X, Toresson G, Benod C, Suh JH, Philips KJ, Webb P, Gustafsson JA (2014) Structure of the retinoid X receptor alpha-liver X receptor beta (RXRalpha-LXRbeta) heterodimer on DNA. Nat Struct Mol Biol 21:277–281
17. Rochel N, Ciesielski F, Godet J, Moman E, Roessle M, Peluso-Iltis C, Moulin M, Haertlein M, Callow P, Mely Y, Svergun DI, Moras D (2011) Common architecture of nuclear receptor heterodimers on DNA direct repeat elements with different spacings. Nat Struct Mol Biol 18:564–570
18. Zhang J, Chalmers MJ, Stayrook KR, Burris LL, Wang Y, Busby SA, Pascal BD, Garcia-Ordonez RD, Bruning JB, Istrate MA, Kojetin DJ, Dodge JA, Burris TP, Griffin PR (2011) DNA binding alters coactivator interaction surfaces of the intact VDR-RXR complex. Nat Struct Mol Biol 18:556–563
19. Chandra V, Huang P, Potluri N, Wu D, Kim Y, Rastinejad F (2013) Multidomain integration in the structure of the HNF-4alpha nuclear receptor complex. Nature 495:394–398
20. Yi P, Wang Z, Feng Q, Pintilie GD, Foulds CE, Lanz RB, Ludtke SJ, Schmid MF, Chiu W, O'Malley BW (2015) Structure of a biologically active estrogen receptor-coactivator complex on DNA. Mol Cell 57:1047–1058

Part II

Nuclear Receptor Agonists and Antagonists

Chapter 2

Lipid Homeostasis and Ligands for Liver X Receptors: Identification and Characterization

Jean-Marc A. Lobaccaro, Claude Beaudoin, Bagora Bayala, Silvère Baron, and Amalia Trousson

Abstract

Screening of *bona fide* ligands for nuclear receptors is a real *tour de force* as the identified molecules are supposed to be able to activate the targeted proteins in cell culture as well as in vivo. Indeed orphan nuclear receptors are putative pharmacologically targets for various diseases. It is thus necessary to have quick and reproductive systems that help in identifying new ligands, agonist or antagonist, before using them in vivo in animal models to check for secondary effects. Here, we describe the transient transfections (homologous and heterologous) used for the screening of ligands for liver X receptor α (LXRα, NR1H3) in HeLa cells.

Key words LXR, Transient transfection, Ligand screening, Pharmacological target, Oxysterols

1 Introduction

Liver X receptors, LXRα (NR1H3) and LXRβ (NR1H2), were discovered in the mid-1990s (for a review see Ref. 1). Although originally identified as pivotal regulators of cholesterol homeostasis [2], the physiological roles of LXRs continue to develop. To date as extensively reviewed [3] there is evidence that both LXRs are involved in lipid and glucose homeostasis, skin development, immunity, neurological functions, inflammation, and cancer. Initially LXRα was described as highly expressed in a restricted subset of tissues known to play an important role in lipid metabolism such as liver, small intestine, and adipose tissue, whereas LXRβ was supposed to be expressed more ubiquitously; however these assertions were based on northern blot analysis and it now seems that almost all tissues express both LXRs, even though at variable levels regarding the studied cells [4].

Historically identified as orphan receptors, Mangelsdorf's group "deorphanized" them [5, 6] by demonstrating that LXRs were bound and activated by cholesterol-derived molecules

Iain J. McEwan (ed.), *The Nuclear Receptor Superfamily: Methods and Protocols*, Methods in Molecular Biology, vol. 1443, DOI 10.1007/978-1-4939-3724-0_2,

known as oxysterols. For the last decade, following phenotype analyses of LXR-deficient mouse models as well as genetic studies in human, LXRs have been at the center of active pharmacological investigations to discover new agonists: deregulation of signaling pathways controlled by these nuclear receptors has been directly associated to numerous metabolic, neurological, and/or cancer diseases (for a review see Ref. 7) and it has been tempting to hypothesize that activating LXRs could prevent and/or treat the associated pathologies.

As members of the nuclear receptor superfamily, LXRs present a common general structure (Fig. 1a) organized with "independent" functional domains [7]. The amino-terminal part of the proteins contains an activating function AF1, which permits the recruitment of ligand-independent co-activators. The DNA-binding domain is located in the center of the protein and is characterized by two zinc fingers, which recognize a core sequence AGGTCAnnnnAGGTCA (Fig. 1b). The carboxy-terminal part of LXRs is composed of a hinge domain that permits the recruitment of corepressors in the absence of ligands and a hydrophobic ligand-binding domain required for dimerization, and a transactivation

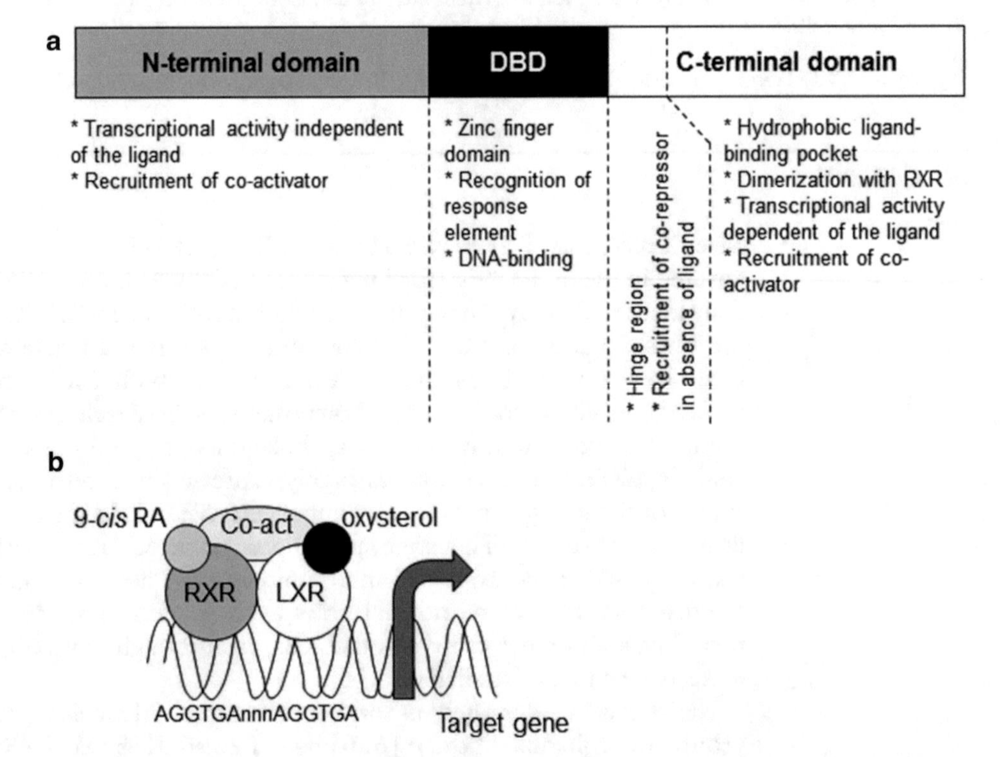

Fig. 1 LXRs are members of the nuclear receptor superfamily. (**a**) Schematic representation of a nuclear receptor. Three main functional domains are usually identified. (**b**) Schematic representation of RXR-LXR heterodimer functioning. When bound by their respective ligands, RXR-LXR recruits co-activators and induces the transcriptional response

domain (AF-2), which recruits co-activators. Bound on DNA as a heterodimer with RXR, the receptor for 9-*cis* retinoic acid, LXRα or LXRβ activates the transcription regulation of its target genes upon ligand binding, which allows the recruitment of co-activators, histone acetyltransferase and RNA polymerase.

Based on these transcriptional properties, screening of LXR ligands is usually based on transient transfection using a reporter gene whose product is easy to detect (here luciferase) and transcription is controlled by an LXR-response element (homologous transfection) [8]. However, in order to exclude any interference of liganded RXR/LXR and the target gene promoter as this is done in mammalian cells, a chimeric system first developed by Brand and Perrimon [9] (heterologous transfection) should be preferred. This UAS/Gal4 system uses upstream activating sequences (5′-CGGRNNRCYNYNCNCCG-3′, UAS) found in *Saccharomyces cerevisiae* galactose-induced genes fused with the reporter gene and a fusion protein made of the GAL4-DNA-binding domain and LXR-ligand-binding domain that can transcribe the target gene could be used (Fig. 2a).

2 Materials

Prepare all solutions using ultrapure water and analytical grade reagents. Prepare and store all reagents at 4 °C (unless indicated otherwise). Diligently follow all waste disposal regulations when disposing waste materials.

2.1 Mammalian Cells

HeLa cells are maintained at 37 °C in an atmosphere of 5% CO_2 with Dulbecco's modified Eagle's medium (DMEM, Life Technologies, St Aubin, France) containing 100 U/ml penicillin and 100 μg/ml streptomycin supplemented with 10% fetal calf serum (Biowest, Nuallié, France).

2.2 Plasmids

For transfection, all plasmids are prepared at a usable final concentration of 100 ng/μl and stored at −20 °C.

1. pCDNA4/T0-Luc [11]: This plasmid allows a constitutive expression of luciferase and serves as a positive control of transfection.
2. UAS-luc: GAL4-responsive MH100(UAS)x4-tk-LUC reporter [10, 11] allows evaluation of the transcriptional activity of human LXRα ligand-binding domain and GAL4-chimera receptor.
3. $_{DBD}$GAL4-$_{LBD}$LXRa: The ligand-binding domain of human LXRα was inserted into pCMX-GAL4 vector to make pCMX-$_{DBD}$GAL4-hLXRα [10, 11].
4. pCMX: The quantity of DNA is maintained constant by addition of empty pCMX vector [10].

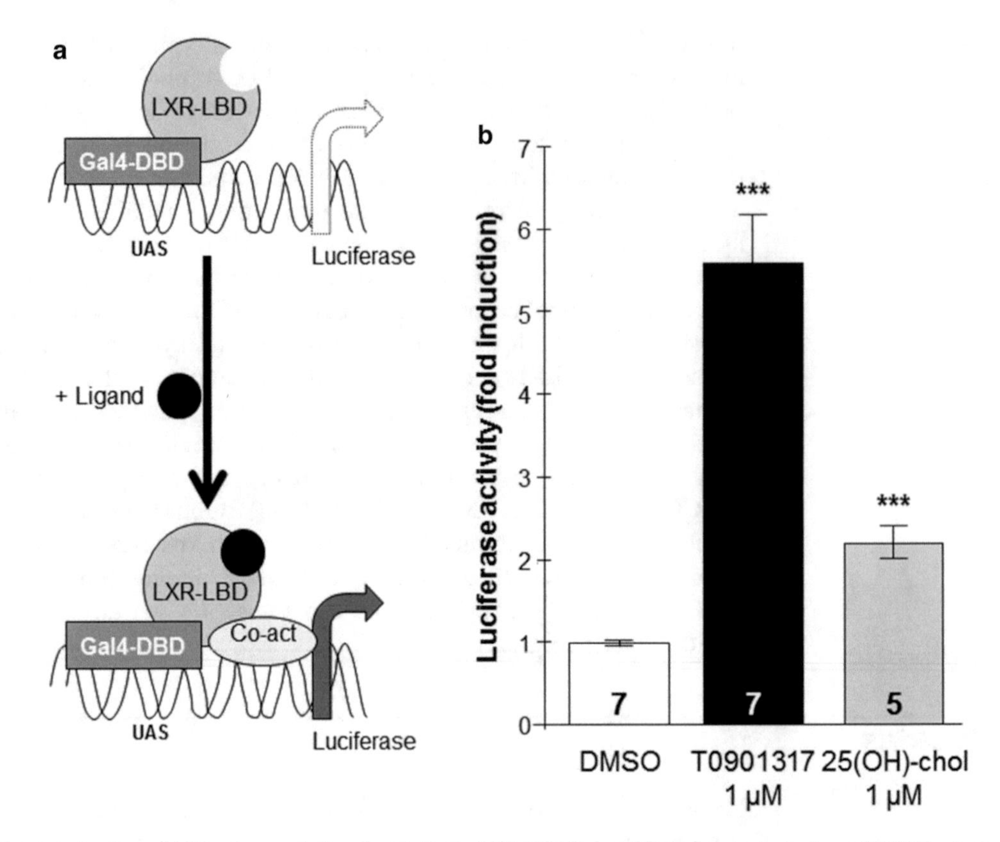

Fig. 2 Induction of LXRα transcriptional activity by T0901317 and 25-hydroxy-cholesterol (25(OH)-chol). (**a**) Schematic representation of heterologous transfection. Gal4-LXR chimeric protein is bound to DNA on upstream activating sequences (UAS). In the absence of ligand, luciferase reporter gene is not induced. In the presence of an agonist, co-activators are recruited and luciferase gene is enhanced. (**b**) Agonistic effects of T0901317 and 25(OH)-chol on human LXRα. Results are indicated as fold induction compared to DMSO. Mean ± SEM. Luciferase activity with pCDNA4/TO-Luc is not shown as the fold induction is usually within the range of 60–300. Number of experiments is indicated in the *histograms*. Student's *t* test: ***$p < 0.0001$

2.3 Ligands

Synthetic ligand T0901317 [12] (Cayman Chemical, Montigny-le-Bretonneux, France) and natural agonist 25-hydroxy-cholesterol [5] (25OH; Sigma Aldrich, L'isle d'Abeau, France) are diluted in DMSO (Sigma Aldrich) and stored at −20 °C at a concentration of 10^{-3} M.

2.4 Specific Reagents for Transfection

1. NaCl 150 mM.
2. Opti-MEM® I Reduced Serum Media (Life Technologies, St Aubin, France) is used as medium during transfection.
3. ExGen 500 in vitro Transfection Reagent (Euromedex, Souffelweyersheim, France) is a cationic polymer transfection reagent used for non-liposomal gene delivery.

2.5 Specific Reagents for Luciferase Detection

1. 5× Reporter lysis buffer (Promega, Charbonnières-les-Bains, France).
2. Luciferase assays are performed in an automated luminometer (see below) with the Genofax A kit (Yelen, Ensue la Redonne, France).

2.6 Apparatus

1. MicroLumatPlus LB96V Microplate Luminometer (Berthold Technologies, Thoiry, France).
2. Vibrating platform shaker Titramax 100 Heidolph (Serlabo Technologies, Entraigues sur la Sorgue, France).

3 Methods

3.1 Cell Transfection

1. On day 0, HeLa cells are seeded at 400×10^3 cells per well in 6-well plates (Dutscher, Brumath, France) in a final volume of 2 ml.
2. On day 1, cells are washed twice with 1 ml of Opti-MEM and then 0.9 ml is added per well (*see* **Note 1**). Plates are then incubated at 37 °C during the plasmid preparation.
3. Plasmids are prepared according to the appropriate combination of plasmids (Table 1) with ExGen 500 (*see* **Note 2**), vortexed for 10 s, and kept under the hood for 10 min. During that time, cells are kept under the hood at room temperature. After 10 min, 100 μl of plasmid solution is added in each well. Plates are gently stirred and then incubated at 37 °C in an atmosphere of 5 % CO_2.

3.2 Induction of Transfection

Six hours after transfection, induction is done by adding 2 ml of DMEM (Life Technologies) with no serum to avoid any presence of endogenous LXR ligand. LXR ligands or vehicle

Table 1
Plasmid combination

Transfection positive control		LXRα experiment	
pCDNA4/T0-Luc	1 μl	UAS-luc $_{DBD}$GAL4-$_{LBD}$LXRα	1 μl 1 μl
pCMX	9 μl	pCMX	8 μl
NaCl 150 mM	40 μl	NaCl 150 mM	40 μl
ExGen 500	4 μl	ExGen 500	4 μl
NaCl 150 mM	46 μl	NaCl 150 mM	46 μl
Total volume	100 μl	Total volume	100 μl

Volumes are indicated in μl per well and plasmids are at a concentration of 100 ng/μl (here T0901317, 25OH and DMSO, respectively) are diluted 1/1000 (*see* **Note 3**) for a final concentration of 10^{-6} M (*see* **Note 4**) in the added DMEM. Cells are then incubated at 37 °C in an atmosphere of 5 % CO_2.

3.3 Development of Luciferase Activity

1. On day 2, 24 h after the induction, cells are washed twice with cold PBS 1× and 100 μl of 1× reporter lysis buffer is added to each well. Plates are rocked on a shaking platform for 10 min at room temperature.
2. The contents of each well are then collected in a 1.5 ml tube and centrifuged at 20,000×*g* for 10 min at 4 °C. 10 μl of cleared supernatant is added to a 96-well plate and then read in a luminometer. 50 μl of luciferase assay solution is automatically added in each well and the signal is recorded for 5 s. An example of results obtained T090317 and 25OH is shown in Fig. 2b (*see* **Note 5**).

4 Notes

1. PBS 1× could be used to wash the cells before adding Opti-MEM. In our experience, more reproducible results are obtained washing the cells with 1 ml of Opti-MEM.
2. Glass tubes must be used for plasmid preparation and ExGen 500 dilution; otherwise, DNA will stick on plastic tube. It is also important to add ExGen 500 in NaCl, never the contrary; likewise for the final mix, ExGen/NaCl solution should be added in the plasmid-containing tube.
3. Many LXR ligands are light sensitive. It is better preparing them under a yellow light and storing them in dark tubes. During addition of ligands in HeLa cells the best would be to work under a yellow light or at least without a direct light.
4. It is important that DMSO is not added above a maximum of 1/1000 to avoid any solvent side effects.
5. Luciferase assay is very sensitive: it is thus important to increase the number of experiments. We routinely perform at least five experiments, each experimental point resulting in eight independent samples.

Acknowledgements

This work was supported by grants from Fondation BNP-Paris, Région Auvergne, Fond Européen de Développement Régional (FEDER), Association de Recherche sur les Tumeurs Prostatiques (ARTP), Fondation ARC, Ligue contre le Cancer for JMAL and

CB, Conférence Episcopale Italienne, Union Economique Monétaire Ouest Africaine (UEMOA), and Campus France for BB. We thank David J. Mangelsdorf (Howard Hughes Medical Institute, Dallas, TX) and M. Makishima (Nihon University School of Medicine, Tokyo, Japan) for plasmids.

References

1. Peet DJ, Janowski BA, Mangelsdorf DJ (1998) The LXRs: a new class of oxysterol receptors. Curr Opin Genet Dev 8:571–575
2. Peet DJ et al (1998) Cholesterol and bile acid metabolism are impaired in mice lacking the nuclear oxysterol receptor LXR alpha. Cell 93:693–704
3. Viennois E et al (2012) Selective liver X receptor modulators (SLiMs): what use in human health? Mol Cell Endocrinol 351:129–141
4. Bookout AL et al (2006) Anatomical profiling of nuclear receptor expression reveals a hierarchical transcriptional network. Cell 126: 789–799
5. Janowski BA, Willy PJ, Devi TR, Falck JR, Mangelsdorf DJ (1996) An oxysterol signalling pathway mediated by the nuclear receptor LXR alpha. Nature 383:728–731
6. Janowski BA et al (1999) Structural requirements of ligands for the oxysterol liver X receptors LXRalpha and LXRbeta. Proc Natl Acad Sci U S A 96:266–271
7. Viennois E et al (2011) Targeting liver X receptors in human health: deadlock or promising trail? Expert Opin Ther Targets 15:219–232
8. Volle DH et al (2004) Regulation of the aldo-keto reductase gene akr1b7 by the nuclear oxysterol receptor LXRalpha (liver X receptor-alpha) in the mouse intestine: putative role of LXRs in lipid detoxification processes. Mol Endocrinol 18:888–898
9. Brand AH, Perrimon N (1993) Targeted gene expression as a means of altering cell fates and generating dominant phenotypes. Development 118:401–415
10. Willy PJ, Mangelsdorf DJ (1997) Unique requirements for retinoid-dependent transcriptional activation by the orphan receptor LXR. Genes Dev 11:289–298
11. Kaneko E et al (2003) Induction of intestinal ATP-binding cassette transporters by a phytosterol-derived liver X receptor agonist. J Biol Chem 278:36091–36098
12. Repa JJ et al (2000) Regulation of mouse sterol regulatory element-binding protein-1c gene (SREBP-1c) by oxysterol receptors, LXRalpha and LXRbeta. Genes Dev 14:2819–2830

Chapter 3

Use of Differential Scanning Fluorimetry to Identify Nuclear Receptor Ligands

Kara A. DeSantis and Jeffrey L. Reinking

Abstract

Identification of small molecules that interact specifically with the ligand-binding domains (LBDs) of nuclear receptors (NRs) can be accomplished using a variety of methodologies. Here, we describe the use of differential scanning fluorimetry to identify these ligands, a technique that requires no modification or derivatization of either the protein or the ligand, and uses an instrument that is becoming increasingly affordable and common in modern molecular biology laboratories, the quantitative, or real-time, PCR machine. Upon being introduced to specific ligands, nuclear receptors undergo structural and dynamic changes that tend to increase molecular stability, which can be measured by the resistance of the protein to heat denaturation. Differential scanning fluorimetry (DSF) uses a dielectric sensitive fluorescent dye to measure the thermal denaturation, or "melting" point (T_m) of a protein under different conditions, in this case in the absence and presence of a candidate ligand. Using DSF, multiple candidates can be screened at once, in numbers corresponding to plate size of the instrument used (e.g., 96- or 384-well), allowing significant throughput if a modest library of compounds needs to be tested.

Key words apo-Ligand-binding domain, Fluorescent dye, Compound library screening

1 Introduction

As liganded NRs are more easily purified, many assays for receptor–ligand interaction involve competition of ligands including newer assays incorporating elements such as fluorescent readout and high-throughput capability [1]. In some cases, highly purified fractions of apo-receptor are not possible, though more stable liganded receptor protein can be purified [2] as it has been proposed that in some cases the LBD is not stable when unbound (as reviewed in Ref. 3). Purification of NR LBD, especially in apo-form, can result in significant loss of initial fraction if further purification steps are undertaken [4, 5]. The changes in conformation and conformational mobility upon ligand binding of NR LBDs [6–11] result in increased molecular stability and an accompanying increased resistance of liganded NR LBD to heat denaturation

Iain J. McEwan (ed.), *The Nuclear Receptor Superfamily: Methods and Protocols*, Methods in Molecular Biology, vol. 1443, DOI 10.1007/978-1-4939-3724-0_3, © Springer Science+Business Media New York 2016

[11–14]. In our hands, differential scanning fluorimetry allows for the use of relatively small quantities of semi-purified fraction of apo-LBD in order to identify ligand binding through changes in the "melting" point of the protein. Since a semi-purified fraction can be utilized, loss of protein as seen in other multi-step purifications [2, 4] can be minimized as only one purification step is necessary.

In the protocol presented herein, we describe the specific steps to express and purify His_6-tagged NR LBDs from recombinant pET15b vectors. During DSF, the fluorescent dye SYPRO Orange is used to monitor the unfolding status of the protein. In order to perform DSF with this dye, the qPCR machine needs to be set to match the excitation/emission profile of the dye. Although DSF can be used to calculate Kd values for ligands with Kds of greater than 10 μM [15], this is well above the range of most NR LBD ligand Kds, so the results obtained from the procedure described here are qualitative in nature, and can be used to determine "hits" within a library of potential ligands.

2 Materials

Prepare all solutions and buffers with ultrapure water; maintain buffers at 4 °C unless otherwise specified. Do not freeze and/or store sample at 4 °C until the following day unless indicated as acceptable at the current step. Break points for overnight freezing or storage at 4 °C will be identified in the procedure. Bleach and dispose of recombinant bacterial solutions as recommended by your facility following proper disposal procedures.

2.1 Bacterial Growth

1. Small batch of LB-amp plates: Add 12.5 g granulated Miller's LB Broth mix (Fisher) and 7.5 g agar to 1 L bottle, add 500 mL of water, and autoclave. Monitor temperature until reduced to 55 °C. Prepare 1000× stock of ampicillin at 100 mg/mL in water, and freeze unused at −20 °C for use in liquid cultures. Add 500 μL ampicillin to a final concentration of 100 μg/mL, and swirl to mix. Pour into petri dishes and allow to cool. Refrigerate once solid.
2. Starter liquid culture: Add 1.25 g of granulated Miller LB Broth mix to 250 mL Erlenmeyer flask, bring up to final volume of 50 mL with water, and autoclave.
3. Large liquid culture: Add 71.4 g of granulated Terrific Broth (Fisher) to a 4 L Erlenmeyer flask, add 1.5 L of water, and autoclave.
4. IPTG solution: Dissolve 2.38 g of Isopropyl-β-d-thiogalactopyranoside (IPTG) in 10 mL of water. Store at −20 °C. Final concentration is 1 M.

2.2 Bacterial Lysis

1. Binding buffer: Dissolve 0.68 g imidazole, 29.2 g NaCl, and 50 mL glycerol in ~800 mL of water. Adjust pH of buffer to 8.0, and then add additional water to a total volume to 1 L. Chill buffer at 4 °C overnight prior to use. Final concentrations in buffer should be as follows: 10 mM imidazole at pH 8.0, 500 mM NaCl, and 5% (v/v) glycerol.
2. Lysis protease inhibition: Add 0.35 g phenylmethanesulfonylfluoride (PMSF) to 10 mL absolute ethanol for a final concentration of 200 mM. Store at −20 °C.
3. Lysozyme: Crystalline egg white lysozyme (Fisher). Store at −20 °C.

2.3 Protein Purification

1. Elution buffer: Dissolve 8.5 g imidazole, 14.6 g NaCl, and 25 mL glycerol in ~400 mL of water. Adjust pH of buffer to 8.0, and then add additional water to a final volume of 500 mL. Chill buffer at 4 °C overnight prior to use. Final concentrations are as follows: 250 mM imidazole at pH 8.0, 500 mM NaCl, and 5% glycerol.
2. Dialysis buffer: Dissolve 26.0 g *N*-(2-hydroxyethyl)piperazine-*N'*-2-ethanesulfonic acid (HEPES), 58.4 g NaCl, 3.08 g dithiothreitol (DTT), and 100 mL glycerol in approximately 1.6 L of water. Adjust pH to 7.5, and then add additional water to a final volume of 2 L. Final concentrations are 50 mM HEPES at pH 7.5, 500 mM NaCl, 10 mM DTT, and 5% glycerol.

2.4 Analysis

1. Analysis buffer: Dissolve 13.0 g HEPES, 29.2 g NaCl, 1.54 g DTT, and 50 mL glycerol in ~800 mL of water. Adjust pH to 7.5, and then add additional water to a total volume of 1 L. Final concentrations are 50 mM HEPES at pH 7.5, 500 mM NaCl, 10 mM DTT, and 5% glycerol.
2. Vehicle solution: Combine 200 μL of ligand solvent (e.g., DMSO or EtOH) with 800 μL of analysis buffer.
3. 10× SYPRO Orange: Prepare a 10× stock of SYPRO orange dye by diluting 2 μL of 5000× SYPRO Orange (Sigma) with 1 mL of analysis buffer. This solution can be stored at −20 ° C.
4. Ligand "master" plate: Candidate ligands should be dissolved in appropriate solvent (e.g., DMSO or EtOH) at 10 mM concentration and stored at −80 ° C. A master plate of the compounds is prepared by thawing candidate compounds and pipetting 2 μL into wells of a 96-well PCR plate. Add 8 μL of analysis buffer to each well, for a final compound concentration of 1 mM. Include one or more wells containing vehicle solution to serve as an unliganded reference. This master plate can be stored at −20 °C after use for several weeks.

3 Methods

3.1 Transformation, Growth, and Induction

1. Using pET-15 plasmid containing the LBD of interest, transform BL-21 (DE3) expression strain *E. coli* (Invitrogen) using heat shock. Plate on LB-amp plates and incubate at 37 °C overnight (*see* **Note 1**).
2. Prepare your starter culture using the 50 mL liquid LB autoclaved previously. Add 50 μL of ampicillin stock after the media has cooled from the autoclave to a final concentration of 100 μg/mL. Select a single colony from the LB agar plate grown from the previous day (*see* **Note 2**). Introduce a single colony to the liquid culture and grow in a shaking incubator at 37 °C overnight.
3. Prepare your protein growth culture (*see* **Note 3**). To the 1.5 L Terrific Broth media previously autoclaved and cooled, add 1.5 mL of ampicillin stock to a final concentration of 100 μg/mL. After addition of ampicillin, withdraw 1 mL of media into a cuvette or tube for a blank for spectrophotometer readings. Cover this temporarily with parafilm, if necessary. Add your starter culture to the 1.5 L culture and incubate, shaking, at 37 °C (*see* **Notes 4** and **5**).
4. Monitor your bacterial growth of your protein growth culture using hourly spectrophotometer readings (*see* **Note 6**) at 600 nm using the reserved media sample as a blank. Induce your culture by adding 375 μL of IPTG (final concentration of 0.25 mM) when your culture reaches an OD_{600} of 0.4–0.8. Drop temperature of incubation to 15 °C for overnight shaking incubation to encourage soluble expression of NR LBD (*see* **Note 7**).

3.2 Lysis and Extraction

1. Pour your 1.5 L culture into centrifuge bottles, centrifuge at 1800 × *g* for 20 min at 4 °C, and pour off supernatant. Resuspend pellet in 30 mL of binding buffer while gently vortexing in order to release pellet. Transfer resuspended cells to 50 mL conical. Keep on ice.
2. While on ice, add 1 mL PMSF and 0.25 mg/mL lysozyme.
3. Cell suspension may be flash frozen and stored at −80 °C at this time (*see* **Note 8**).
4. Sonicate for three to four 2-min rounds (*see* **Note 9**). Transfer lysate to high-speed centrifuge tubes.
5. Centrifuge lysate at 30,000 × *g* at 4 °C for 30 min. Transfer supernatant to 50 mL conicals, and keep on ice.

3.3 Protein Purification (Ideally Performed at 4 °C)

1. Pack column using Ni-NTA Superflow (Qiagen) beads. If using 10 mL gravity column, use 2 mL of beads. Alternatively, FPLC can also be used for the purification using the same resin (*see* **Note 10**). Equilibrate column with one column volume of cold binding buffer prior to introducing lysate. Slowly pass lysate through column. Using gravity column, make sure that lysate passes through at a rate of visible singular drips.
2. Wash column with at least 300 mL of cold binding buffer.
3. Elute sample with 1 mL fractions of elution buffer. Test for the presence of protein using Bradford reagent (Bio-Rad) and pool fractions with significant protein content.
4. Introduce pooled protein samples into dialysis tubing, and dialyze overnight in dialysis buffer.
5. Verify the presence of LBD in sample using SDS-PAGE based on expected molecular weight (Fig. 1) (*see* **Note 11**). Quantify protein in sample using Bradford Assay or UV absorption.

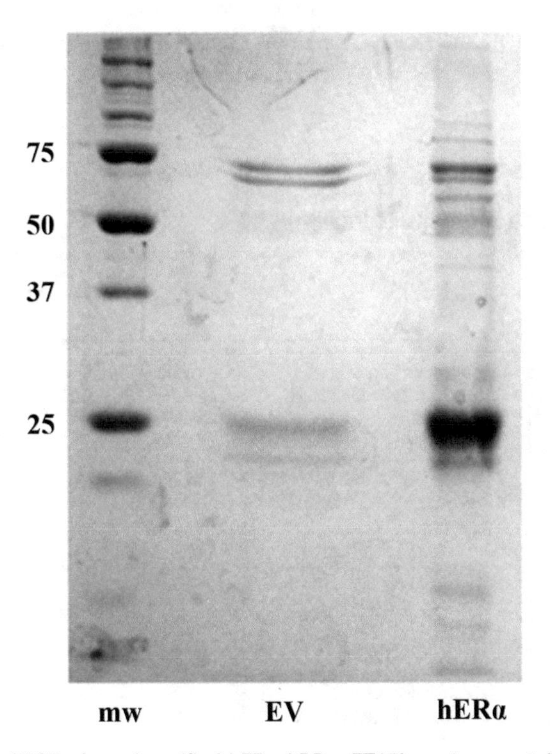

Fig. 1 SDS-PAGE of semi-purified hERα LBD. pET15b vectors containing either human ER LBD (302–355) (kindly provided by Dino Moras, Ref. 17) or no insert ("empty" Vector, EV) were expressed and purified using the protocols detailed herein. Purifications in this gel were derived from 6 L of bacterial culture. The lane for hERα was loaded with 50 μg of protein as determined by Bradford assay. An equivalent volume of eluent was used for the EV lane. Protein purity shown is typical of purity able to be used in assay

3.4 Determining Assay Conditions

1. In order to determine the optimal protein and SYPRO Orange concentrations for your preparation of protein, combine ingredients in individual wells of a PCR plate as depicted in Table 1 (*see* **Note 12**).
2. Pipet up and down several times when adding the final ingredient in order to mix the contents (*see* **Note 13**).
3. Cover plate with optically clear film.
4. Centrifuge plate at 2000×*g* for 10 min to remove air bubbles.
5. Using a qPCR machine with excitation and emission set to match SYPRO Orange and programmed to increase temperature of samples from 25 °C to 95 °C at a rate of 1 °C/min, take fluorescence measurement every 0.2 °C (*see* **Note 14**).
6. Examine the resultant denaturation curves. A typical curve will show an increase in relative fluorescence units (RFUs) as the protein denatures (e.g., Fig. 2a). At high temperatures, the RFU values drop, presumably due to denatured protein forming insoluble precipitates and no longer interacting with the SYPRO dye. Choose the condition with the greatest magnitude of the increase in RFU in the first part of the curve. If multiple

Table 1
Determination of assay conditions

[Protein]	0.5X	1X	2X	[SYPRO Orange]
	2 μL "vehicle" solution 0.5 μL 10X SYPRO Orange 1 μL protein 6.5 μL Analysis Buffer	2 μL "vehicle" solution 1 μL 10X SYPRO Orange 1 μL protein 6 μL Analysis Buffer	2 μL "vehicle" solution 2 μL 10X SYPRO Orange 1 μL protein 5 μL Analysis Buffer	1μL
	2 μL "vehicle" solution 0.5 μL 10X SYPRO Orange 2 μL protein 5.5 μL Analysis Buffer	2 μL "vehicle" solution 1 μL 10X SYPRO Orange 2 μL protein 5 μL Analysis Buffer	2 μL "vehicle" solution 2 μL 10X SYPRO Orange 2 μL protein 4 μL Analysis Buffer	2μL
	2 μL "vehicle" solution 0.5 μL 10X SYPRO Orange 4 μL protein 3.5 μL Analysis Buffer	2 μL "vehicle" solution 1 μL 10X SYPRO Orange 4 μL protein 3 μL Analysis Buffer	2 μL "vehicle" solution 2 μL 10X SYPRO Orange 4 μL protein 2 μL Analysis Buffer	4μL
	2 μL "vehicle" solution 0.5 μL 10X SYPRO Orange 6 μL protein 1.5 μL Analysis Buffer	2 μL "vehicle" solution 1 μL 10X SYPRO Orange 6 μL protein 1 μL Analysis Buffer	2 μL "vehicle" solution 2 μL 10X SYPRO Orange 6 μL protein 0 μL Analysis Buffer	6μL

conditions have similar ΔRFU values, choose conditions that minimize protein usage (*see* **Note 15**).

7. This process of "gridding out" assay conditions should be repeated for each new protein preparation.

3.5 Screening Compounds

1. Combine 2 μL of component well of the ligand "master" plate in a well of an experimental plate with volumes of protein, 10× SYPRO Orange, and analysis buffer as determined by the assay from the previous section (*see* **Note 16**).
2. Repeat **steps 2–5** from the previous section.
3. When analyzing results, use the first derivative plot (Fig. 2b, d) to determine the T_m of the protein for that particular condition.
4. Ligand "hits" are those which significantly increase the T_m of the protein as compared to the vehicle controls (Fig. 3). The magnitude of ΔT_m is negatively correlated to Kd of the interaction [14].

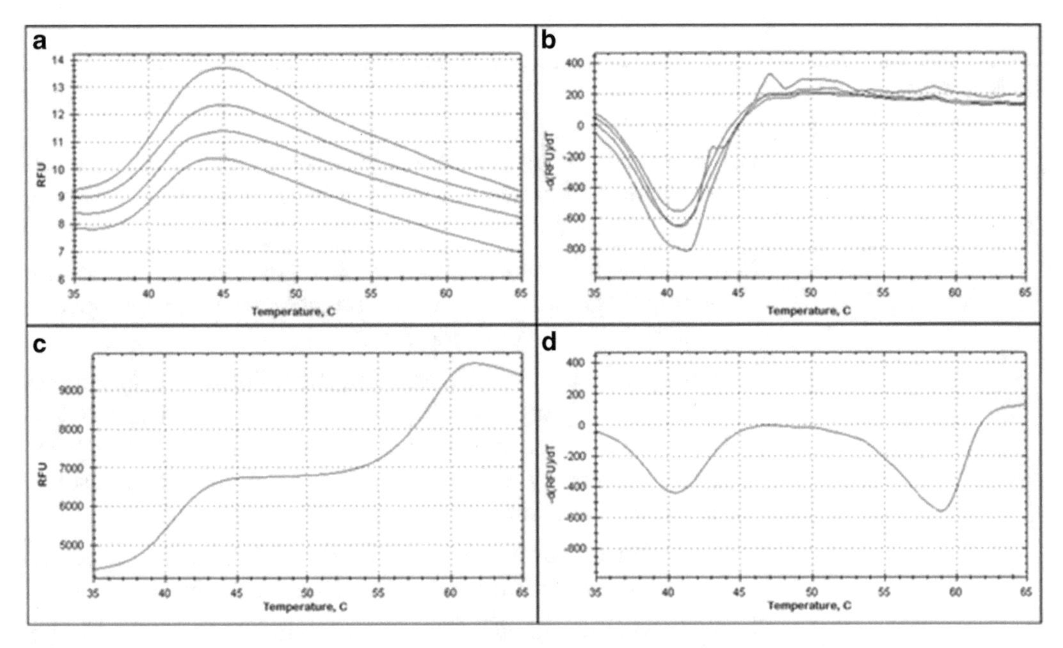

Fig. 2 Use of DSF to determine protein T_m. Four replicate vehicle wells were included in the master ligand plate, representing unliganded melting curves of hERα LBD (**a**). The first derivative (as calculated by instrument software) of the curves allows determination of the T_m of unliganded hER α LBD at the inverse peak (**b**). The melting curve (**c**) and first derivative (**d**) of the well containing estradiol, a high-affinity natural ligand of hERα LBD, are also shown. When multiple inverse peaks are present, such as in panel (**d**), the peak with the higher T_m is considered. The lower peak can be attributed to non-binding conformations of the NR LBD and/or contaminants present in the preparation

Fig. 3 Use of DSF to identify "hits" in a compound library. hERα LBD was screened against the 76-compound Screen-Well Nuclear Receptor ligand library (Enzo Life Sciences, BML-2802). The *bar graph* represents the average T_m obtained from each condition, performed in triplicate. *Error bars* represent the standard error of the mean. The first four components of the graph are the four vehicle wells that represent unliganded hER α LBD. "Hits" from the library are indicated by *arrows*, and correspond to tamoxifen, estrone, and estradiol, from *left to right*

4 Notes

1. We sometimes use carbenicillin in place of ampicillin in the preparation of plates since it is more heat stable and leads to fewer satellite colonies.
2. Avoid satellite colony selection. If plate must be stored after overnight growth, it may be possible to store wrapped in parafilm at 4 °C for several days.
3. In place of pre-mixed LB and TB, the following recipes can be used. LB: 10 g tryptone, 5 g yeast extract, and 10 g NaCl per L of water. TB: 12 g tryptone, 24 g yeast extract, 2.2 g potassium phosphate monobasic, and 9.4 g potassium phosphate dibasic per L of water.
4. If your starter culture does not look visibly turbid, re-inoculate another starter culture and begin from overnight culture again.
5. We have also had success using auto-inducing media [16] to produce NR LBDs. Grow the large-scale culture in ZYP-5052 for approximately 4–6 h at 37 °C until the flask is visibly turbid, and then reduce the temperature to 15 °C for overnight induction.
6. More frequent readings may be necessary when log phase of bacterial growth is reached.
7. If you do not have the capacity for shaking incubation at 15 °C, shaking at room temp or on ice may be possible.

8. In order to thaw frozen cell suspensions, place cells under running cold water. The freeze/thaw action will help to lyse cells, so 1 mL of fresh PMSF solution needs to be present.
9. Take care not to heat or foam the sample during sonication. Sonicate while on ice if possible.
10. When using an FPLC instead of gravity flow, we recommend substituting 3 % v/v propylene glycol for 5 % v/v glycerol in all solutions to reduce viscosity.
11. For this assay, the protein does not need to be purified to homogeneity.
12. If measured protein concentration is above 10 mg/mL, consider dilution of the protein prior to setting up the condition assay grid.
13. In our hands, excessive mixing, such as by use of plate shaker, leads to degradation of results, presumably due to protein precipitation.
14. The qPCR program can be significantly shortened by ramping only between 25 and 75 °C, increasing the rate of temperature increase to 2 °C/min and less frequent fluorescence readings. Although this can reduce the precision of the melting point determined, the optimal conditions for initial screening can be readily obtained under these accelerated conditions.
15. In our hands, some nuclear receptor LBDs do not produce curves in the apo form as described in the protocol that cannot be used to determine a T_m. In these cases, use a ligand known to bind to the LBD (if available) as a positive control in place of the vehicle solution in the "gridding out" protocol to determine optimal assay conditions. The ligand should be prepared in the same way as components of the ligand "master" plate. "Hits" are candidate ligands that produce similar to the positive control.
16. Use of multichannel pipettors or a robotic liquid handler can expedite preparation of the experimental PCR plate. Use the same format of plate to prepare the ligand "master" plate as the experimental plate. The SYPRO Orange, protein, and analysis buffer solutions can be pipetted from reagent troughs.

References

1. Parker G, Law T, Lenoch F et al (2000) Development of high throughput screening assays using fluorescence polarization: nuclear receptor–ligand-binding and kinase phosphatase assays. J Biomol Screen 5:77–88
2. Juzumiene D, Chang C, Fan D et al (2005) Single-step purification of full-length human androgen receptor. Nucl Recept Signal 3:1–5
3. Ingraham H, Redinbo M (2005) Orphan nuclear receptors adopted by crystallography. Curr Opin Struct Biol 15:708–715
4. Razzera G, Vernal J, Portugal RV et al (2004) Expression, purification, and initial structural characterization of rat orphan nuclear receptor NOR-1 LBD domain. Protein Expr Purif 37:443–449

5. Graham L, Pilling P, Eaton R et al (2007) Purification and characterization of recombinant ligand binding domains from the ecdysone receptors of four pest insects. Protein Expr Purif 53:309–324
6. Schupp M, Lazar M (2010) Endogenous ligands for nuclear receptors: digging deeper. J Biol Chem 285:40409–40415
7. Allan G, Leng X, Tsai S et al (1992) Hormone and antihormone induce distinct conformational changes which are central to steroid receptor activation. J Biol Chem 267:19513–19520
8. Hamuro Y, Coales SJ, Morrow JA et al (2006) Hydrogen/deuterium-exchange (H/D-Ex of PPARgamma LBD in the presence of various modulators. Protein Sci 15:1883–1892
9. Johnson BA, Wilson EM, Li Y et al (2000) Ligand-induced stabilization of PPAR gamma monitored by NMR spectroscopy: implications for nuclear receptor activation. J Mol Biol 298:187–194
10. Yan X, Broderick D, Leid ME et al (2004) Dynamics and ligand-induced solvent accessibility changes in human retinoid X receptor homodimer determined by hydrogen deuterium exchange and mass spectrometry. Biochemistry 43:909–917
11. Toney J, Wu L, Summerfield A et al (1993) Conformational changes in chicken thyroid hormone receptor alpha 1 induced by binding to ligand or to DNA. Biochemistry 32:2–6
12. Pissios P, Tzameli I, Kushner P et al (2000) Dynamic stabilization of nuclear receptor. Mol Cell 6:245–253
13. Reinking J, Lam MM, Pardee K et al (2005) The Drosophila nuclear receptor e75 contains heme and is gas responsive. Cell 122: 195–207
14. DeSantis K, Reed A, Rahhal R et al (2012) Use of differential scanning fluorimetry as a high-throughput assay to identify nuclear receptor ligands. Nucl Recept Signal 10:e002
15. Vivoli M, Ayres E, Beaumont E et al (2014) Structural insights into Wcbl, a novel polysaccharide-biosynthesis enzyme. IUCrJ 1:28–38
16. Studier W (2005) Protein production by auto-induction in high-density shaking cultures. Protein Expr Purif 41:207–234
17. Eiler S, Gangloff M, Duclaud S et al (2001) Overexpression, purification, and crystal structure of native ER alpha LBD. Protein Expr Purif 22:165–173

Chapter 4

Drug-Discovery Pipeline for Novel Inhibitors of the Androgen Receptor

Kush Dalal*, Ravi Munuganti*, Hélène Morin, Nada Lallous, Paul S. Rennie*, and Artem Cherkasov*

Abstract

The androgen receptor (AR) is an important regulator of genes responsible for the development and recurrence of prostate cancer. Current therapies for this disease rely on small-molecule inhibitors that block the transcriptional activity of the AR. Recently, major advances in the development of novel AR inhibitors resulted from X-ray crystallographic information on the receptor and utilization of in silico drug design synergized with rigorous experimental testing.

Herein, we describe a drug-discovery pipeline for in silico screening for small molecules that target an allosteric region on the AR termed the binding-function 3 (BF3) site. Following the identification of potential candidates, the compounds are tested in cell culture and biochemical assays for their ability to interact with and inhibit the AR. The described pipeline is readily accessible and could be applied in drug design efforts toward any surface-exposed region on the AR or other related steroid nuclear receptor.

Key words Androgen receptor, Prostate cancer, Virtual screening, Computer-aided drug design

1 Introduction

The androgen receptor (AR) is a ligand-activated transcription factor that contributes to the growth and recurrence of prostate cancer [1, 2]. The AR contains three major structural parts: an N-terminal domain (NTD), followed by a DNA-binding (DBD) and a ligand-binding (LBD) domain [3]. The receptor is activated by androgens, such as dihydrotestosterone (DHT), which bind to the LBD and cause the transcription factor to dimerize, enter the nucleus, and drive the expression of target genes to promote tumor growth [4, 5]. The current therapies for prostate cancer employ small-molecule inhibitors (i.e., anti-androgens and AR-antagonists) to compete with DHT for binding to the LBD, thus preventing AR activation [6]. The creation of better AR-LBD-interacting

* Author contributed equally with all other contributors.

Iain J. McEwan (ed.), *The Nuclear Receptor Superfamily: Methods and Protocols*, Methods in Molecular Biology, vol. 1443, DOI 10.1007/978-1-4939-3724-0_4, © Springer Science+Business Media New York 2016

compounds involves using known crystallographic information on this domain [7–9] to model potential inhibitors into the androgen-binding site (ABS), or other functional regions including the activation-function 2 (AF2) and binding-function 3 (BF3), which both serve to recruit cofactors [10–13].

Herein we describe our established in silico drug-design pipeline [12, 14–17] enabling virtual screening of an electronic library of compounds to find candidates capable of selective binding to the BF3 region. This approach represents a rational way of selecting promising candidates from a vast library of chemicals with respect to their potentially favorable characteristics such as binding affinity to the target, potency, toxicity, and chemical integrity. Potential candidates are then experimentally validated using a battery of assays. LNCaP cells bearing a stably incorporated enhanced green fluorescent protein (eGFP) fluorescent reporter are used to determine if candidate compounds can prevent AR transcriptional activity. Hit compounds are further tested to establish their dose-dependent inhibition curves and the corresponding IC_{50} values. Finally, the recombinant AR-LBD protein is purified and its direct interaction with small molecules is measured using biolayer interferometry (BLI). Together, these methods allow the discovery and characterization of new small-molecule AR inhibitors. A similar strategy could be applied toward the domains of other steroid nuclear receptors if crystallographic information is available.

2 Materials

2.1 Materials to Perform In Silico Studies

1. Access to www.rcsb.org.
2. ZINC lead-like database: A free database of commercially available compounds for virtual screening (zinc.docking.org). It contains ~4 million purchasable druglike compounds in ready-to-dock, 3D formats. ZINC is provided by the Shoichet Laboratory at the University of California, San Francisco (UCSF), USA.
3. Maestro Suite 9.0., Schrodinger Inc. (Portland, OR, USA).
4. MOE 2012, Chemical Computing Group Inc. (Montreal, QC, Canada).
5. eHiTS 2012, Simbiosys Inc. (Toronto, ON, Canada).
6. A computer cluster composed of 2128 CPU cores (Intel Xeon 2.1 GHz) with a total of 400 TB of central hard drives. The cluster is purchased from IBM and located in Vancouver Prostate Centre, Vancouver, Canada (private access).
7. A desktop computer with Intel (R) Core (TM) i5 CPU at 3.20 GHz processor, 16 GB RAM and nVIDIA Quadro 600 graphics including 3D Vision Pro, Windows 7 (64-bit OS).

2.2 Single Concentration Screening of Compounds for Transcriptional Inhibition of the AR in Cell Culture

1. LNCaP cells bearing a stably transfected eGFP reporter (LNCaP-eGFP) under the control of a probasin promoter (androgen responsive region, ARR-2 PB) [12] (*see* **Note 1**).
2. Phenol-red-free RPMI (Life Technologies) + 5% charcoal-stripped serum (CSS; Thermo Scientific) preheated at 37 °C (RPMI + CSS).
3. Trypsin +0.25% EDTA (Thermo Scientific).
4. Methyltrienolone synthetic androgen (R1881) at 1 μM and 100 nM concentration in 100% ethanol.
5. Bicalutamide-positive control at 100 mM concentration in 100% DMSO.
6. Candidate compounds identified by virtual screening at 10 or 50 mM in 100% DMSO.
7. Sterile 10 cm treated cell culture dish (Corning).
8. Sterile 1.75 mL plastic centrifuge tubes (Axygen).
9. Sterile 24-well clear plate (Corning).
10. Sterile 96-well clear plate (Corning).
11. Sterile 96-well black plate with clear glass bottom (Corning).
12. Sterile 25 mL reservoir (Corning).
13. 50 mL Centrifuge tubes (Corning).
14. 8-Channel multichannel pipet (Gilson).
15. Bio-Rad TC10 Automated Cell Counter and dual-chamber slides (Bio-Rad).
16. Incubator, 37 °C, 5% CO_2.

2.3 Dose-Dependent Inhibition of Promising Compounds

1. Items from Subheading 2.2.
2. 1× Phosphate-buffered saline.
3. Cobas e411 analyzer and measurement vials.

2.4 Purification of the Androgen Receptor Ligand-Binding Domain

1. Strain *Escherichia coli* BL21 (DE3) co-transformed with (a) plasmid Pan4 (avidity) encoding the AR-LBD (amino acids) with N-terminal AviTag sequence (GLNDIFEAQKIEWHE) and 6-histidine tag (ampicillin resistance), and (b) plasmid pBirAcm encoding for biotin ligase enzyme (chloramphenicol resistance).
2. Luria Bertani (LB) broth (Sigma).
3. EDTA-free protease inhibitor cocktail tablets (complete, Roche).
4. 5 mM Biotin dissolved in 10 mM bicine pH 8.3.
5. 100 mM Dihydrotestosterone (DHT) in 100% ethanol (*see* **Note 2**).

6. 1 M Isopropyl β-D-1-thiogalactopyranoside (IPTG) (Sigma).
7. Lysis buffer: 50 mM HEPES pH 7.5, 150 mM Li_2SO_4, 10 mM imidazole, 10% glycerol, 0.5 mM TCEP, and 20 μM DHT.
8. Wash buffer: 50 mM HEPES pH 7.5, 150 mM Li_2SO_4, 20 mM imidazole.
9. Elution buffer: 50 mM HEPES pH 7.5, 150 mM Li_2SO_4, 250 mM imidazole.
10. Dilution buffer: 50 mM HEPES pH 7.5, 150 mM Li_2SO_4, 10% glycerol, 0.5 mM TCEP, and 20 μM DHT.
11. Sonicator: 550 Sonic Dismembrator (Fisher Scientific).
12. Ni-NTA agarose metal affinity beads (GE Healthcare, Qiagen).
13. Poly-prep 20 mL gravity chromatography columns (Bio-Rad).
14. Sodium dodecyl sulfate (SDS)-polyacrylamide gel electrophoresis (PAGE) buffers and apparatus for a running a 10% SDS-PAGE gel: 5× Sample buffer (250 mM Tris–HCl pH 7.0, 10% (w/v) SDS, 10% (v/v) β-mercaptoethanol (BME), 50% glycerol, 0.05% bromophenol blue); 1.5 M Tris–HCl pH 8.8; 0.5 M Tris–HCl pH 6.8; 10% ammonium persulfate (APS) in MilliQ H2O; 40% mixed acrylamide/bis solution (37:5:1) (Bio-Rad); *N,N,N,N′*-tetramethyl-ethylenediamine (TEMED) (Bio-Rad); 10× SDS-PAGE electrophoresis running buffer (250 mM Tris base, 1.9 M glycine, 0.15% (w/v) SDS).
15. Amicon 10 kDa MWCO centrifugal filter units (Amicon Ultra, Millipore).
16. 10 mL Zeba Spin Desalting Column, 7 kDa MWCO (Thermo Scientific).

2.5 Biolayer Interferometry Screening of AR-LBD-Interacting Compounds

1. ForteBIO Octet Red Biolayer Interferometry instrument (Menlow Park, California).
2. Super Streptavidin (SSA) Dip and Read Biosensor needles (ForteBIO).
3. BLI buffer (150 mM Li_2SO_4, 50 mM HEPES pH 7.0, 1 mM dithiothreitol [DTT], 10 mM DHT).
4. Plate accessory consisting of green tray and lid (ForteBIO).
5. Black 96-well plate (Grenier Bio).
6. Biotinylated purified AR-LBD at approximately 1 g/L concentration.
7. Biocytin (biotinyl-L-lysine) 10 g/L stock.
8. SuperBlock Blocking Buffer in TBS (Pierce, Thermo Scientific).

3 Methods

The developed drug-discovery pipeline consists of two major parts: (1) computational modeling of the BF3 pocket and virtual screening of a library of compounds against it, and (2) cell culture-based and in vitro-based assays with candidate compounds that are predicted to inhibit the AR transcriptional activity by physically interacting with the AR BF3.

In preparation for virtual screening of compounds, the AR BF3 crystal structure must be first processed with Maestro suite, an all-purpose molecular modeling environment used to compute the structural features of the targeted binding interface. By using this tool, a receptor grid defining the physical and chemical characteristics of the BF3 interface can be established (Fig. 1a, b). The subsequent screening of the ZINC database (four million compounds) involves "funneling" compounds through a series of scoring functions and manual inspections of the estimated docking poses to eliminate most compounds and enrich the dataset for others with favorable characteristics. The structure of the AR-LBD is first prepared for docking simulations by "washing" the structure to remove unwanted and extraneous atoms such as water and salt (Fig. 2). Docking of compound to the BF3 site is consequently performed with the Glide SP program (Fig. 3), followed by the analysis of candidate compounds with the more stringent electronic high-throughput screening (eHiTS) docking protocol. In the next step, surviving compounds are analyzed by the MOE program (Molecular Operating Environment) to determine the most energetically favorable docking poses for each candidate compound by calculating the root mean squared deviation between Glide and eHiTS output (Fig. 4). The further refinement of hydrogen bond energies, hydrophobic interactions, and receptor/ligand flexibility is performed to characterize and exclude undesirable compounds. Finally, all the criteria including docking scores, RMSD values, and refinement scores are used to rank the retained compounds, creating a master list for final visual inspection. As the result, the most promising purchasable compounds are selected and obtained.

The experimental evaluation of the selected substances begins with testing their ability to inhibit the AR transcriptional activity using the eGFP reporter construct in LNCaP prostate cancer cells (Fig. 5a). Potent inhibitory compounds are then tested over a range of concentrations to determine dose-dependent inhibition of eGFP expression (Fig. 5b) and, similarly, for the ability to decrease the level of secreted prostate-specific antigen (PSA) (Fig. 5c), a naturally occurring AR-regulated gene.

In parallel, to determine if candidate compounds can bind to the predicted target site, a biotinylated form of the AR-LBD must

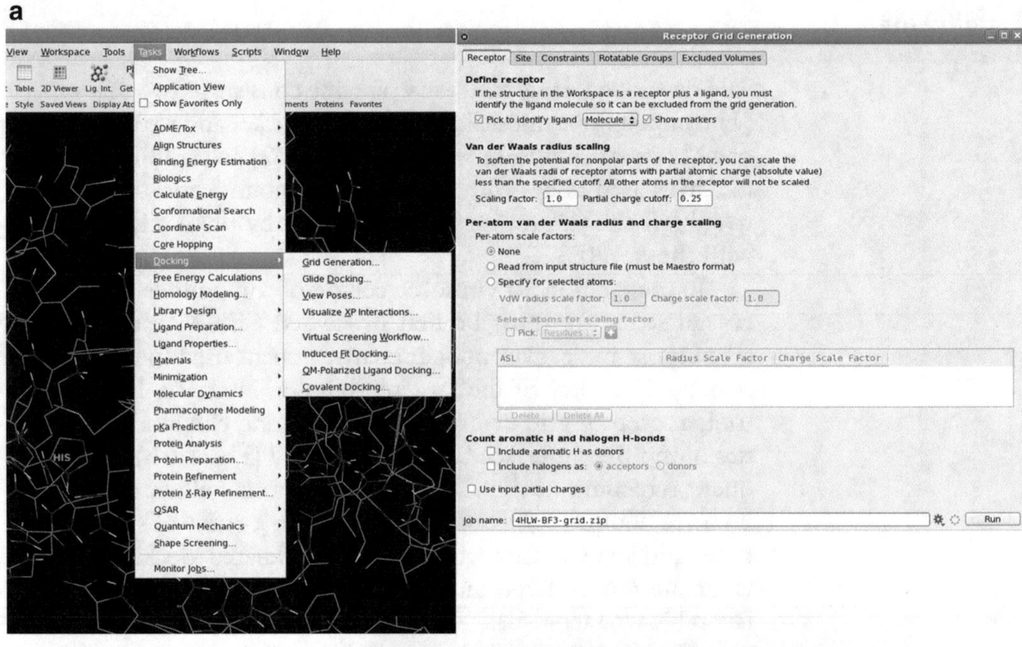

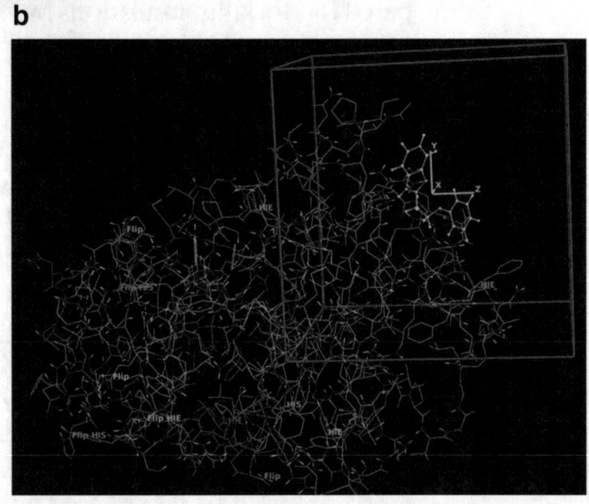

Fig. 1 (**a**) Maestro panels illustrating the receptor grid generation procedure. (**b**) Crystal structure of androgen receptor (4HLW). 17 W, a crystal ligand bound to BF3 pocket is shown in *green*. Grid generated based on 17 W is shown in *pink*

first be expressed in *E. coli* and then purified by metal affinity chromatography (Fig. 5d). This purified product is required for biophysical characterization by biolayer interferometry, a technique used to measure a direct interaction between the compounds and the AR-LBD (Fig. 5e) (*see* **Note 3**).

3.1 Preparation of the 3D Structure of the AR LBD

1. The virtual screening was carried out on the AR-LBD crystal structure (PDB code 4HLW, 1.80 Å resolution). A typical PDB structure file consists only of heavy atoms, water molecules, and some hetero atoms such as chloride and sulfate ions.

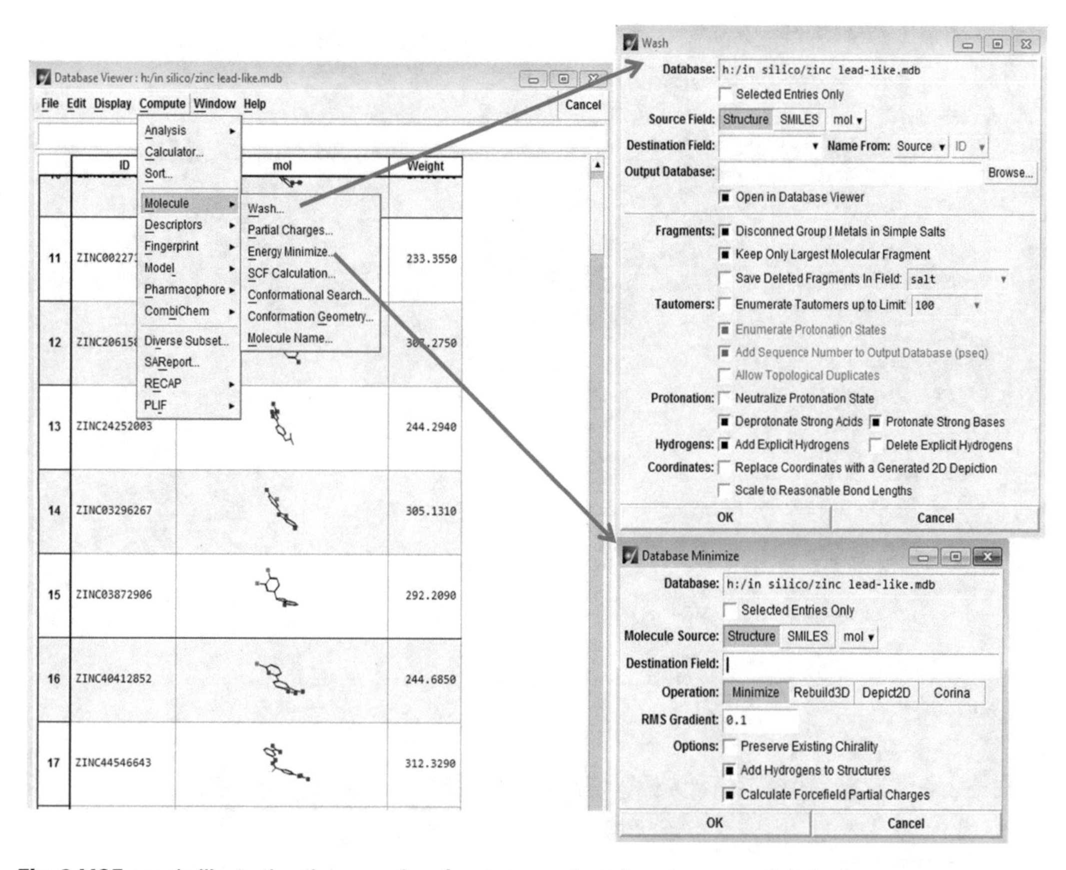

Fig. 2 MOE panels illustrating the procedure for compound wash and energy minimization

The structure generally has no information on bond orders, surface topology, or formal atomic charges. Hence, we must prepare the protein structure accordingly to predict these characteristics (*see* **Note 4**).

2. 4HLW is downloaded from www.rcsb.org. The file is saved as 4HLW.pdb.
3. Import the pdb file into Maestro 9.0 suite (www.schrodinger.com). The original 4HLW PDB file contains glycerol, sulfate ion, waters, testosterone (**TES**), and 2-[(2-phenoxyethyl) sulfanyl]-1H-benzimidazole (**17 W**) as part of the crystal structure. Heteroatoms which are not required for the docking of compounds (glycerol, sulfate ion, and waters in this particular case) are deleted using *Maestro> Protein Preparation Wizard | Review and Modify* in Maestro suite.
4. Bond order for the crystal ligands (**TES** and **17 W**) and protein structure are adjusted using *Maestro> Protein Preparation Wizard | Import and Process.*
5. Using *Maestro>| Protein Preparation Wizard | Refine,* the missing hydrogen atoms are added to the protein structure.

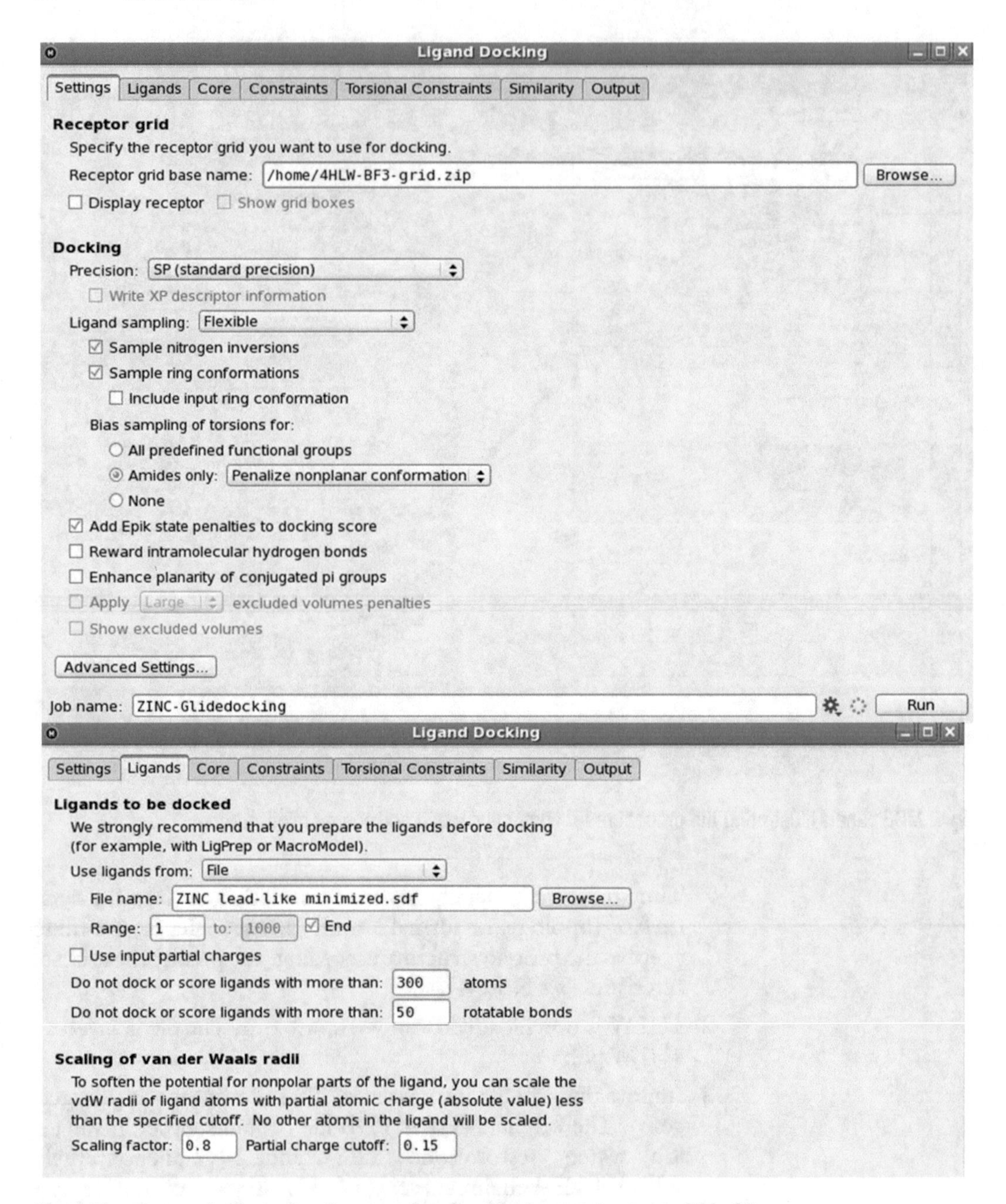

Fig. 3 Maestro panels illustrating the procedure for virtual screening using Glide SP program

In addition, only amino acid side chains of the crystal structure are energy-minimized using the OPLS_2005 force field, as implemented by Maestro with default settings (the backbone fragments are kept "frozen").

3.2 Receptor Grid Generation

1. A receptor grid file represents physical properties of a binding pocket of the receptor that are searched when attempting to dock a compound.

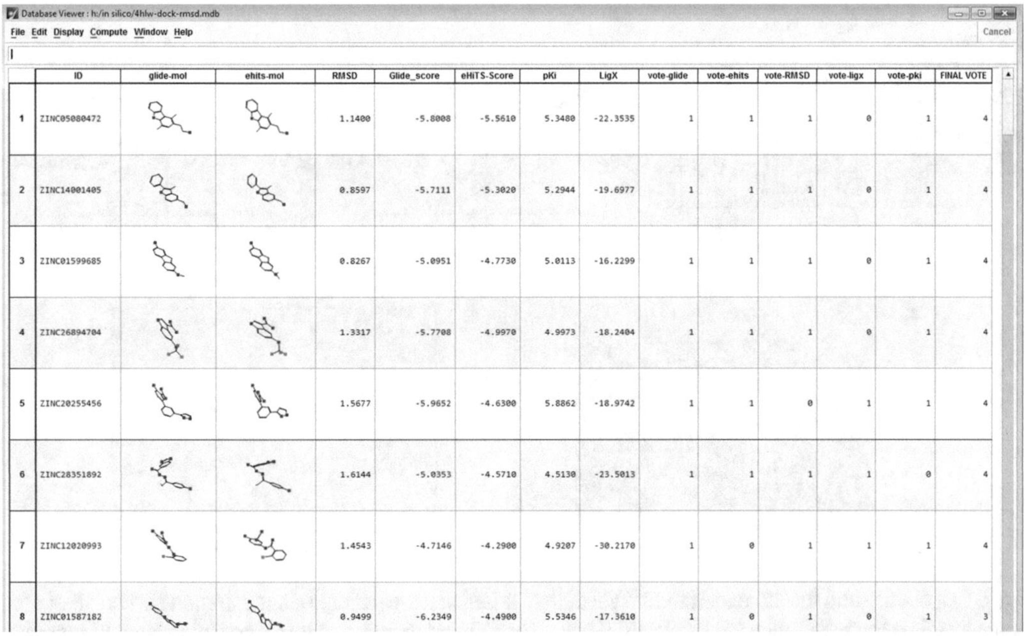

Database Viewer : h:/in silico/4hlw-dock-rmsd.mdb

File Edit Display Compute Window Help

Cancel

	ID	glide-mol	ehits-mol	RMSD	Glide_score	eHiTS-Score	pKi	LigX	vote-glide	vote-ehits	vote-RMSD	vote-ligx	vote-pki	FINAL VOTE
1	ZINC05080472			1.1400	-5.8008	-5.5610	5.3480	-22.3535	1	1	1	0	1	4
2	ZINC14001405			0.8597	-5.7111	-5.3020	5.2944	-19.6977	1	1	1	0	1	4
3	ZINC01599685			0.8267	-5.0951	-4.7730	5.0113	-16.2299	1	1	1	0	1	4
4	ZINC26894704			1.3317	-5.7708	-4.9970	4.9973	-18.2404	1	1	1	0	1	4
5	ZINC20255456			1.5677	-5.9652	-4.6300	5.8862	-18.9742	1	1	0	1	1	4
6	ZINC28351892			1.6144	-5.0353	-4.5710	4.5130	-23.5013	1	1	1	1	0	4
7	ZINC12020993			1.4543	-4.7146	-4.2900	4.9207	-30.2170	1	0	1	1	1	4
8	ZINC01587182			0.9499	-6.2349	-4.4900	5.5346	-17.3610	1	0	1	0	1	3

Fig. 4 An example of MOE database file containing docking conformations from Glide SP and eHITS. Voting procedure is also shown

2. There are three binding sites on the AR ligand-binding domain: (a) the hormone-binding pocket where **TES** is bound; (b) activation function-2 site (AF2); and (c) binding function-3 pocket (BF3), where **17 W** is bound. In this example, a receptor grid for BF3 pocket is generated using **17 W**.
3. Go to *Maestro> Tasks | Docking | Grid generation*. Click the tab *Receptor* and select the ligand **17 W** from the workspace. Dark green markers appear on the 17 W. In the van der Waals radii scaling section, ensure that *Scaling factor* is set to the default value of 1.00 (no scaling) (Fig. 1a).
4. Click the *Site* tab. A binding region defined by a 10 Å*10 Å*10 Å box centered on the 17 W is generated (Fig. 1b). Ensure that the *Center* option selected is *Centroid of Workspace ligand.* Use default settings for all other adjustable parameters.
5. Change the *job name* to "4HLW-BF3-grid.zip" and click *Run*. It may take 5–10 min to create the grid file.

3.3 Ligand Preparation for Virtual Screening

1. The ZINC lead-like database is used for virtual screening. This database was downloaded from zinc.docking.org in structure-data file (SDF) format. SD files are ASCII text files that adhere to a strict format for representing multiple chemical structure records and associated data fields (such as molecular weight).
2. The compounds in SD file are imported into an MOE Database Viewer (.mdb) format using MOE version 2012. The MOE

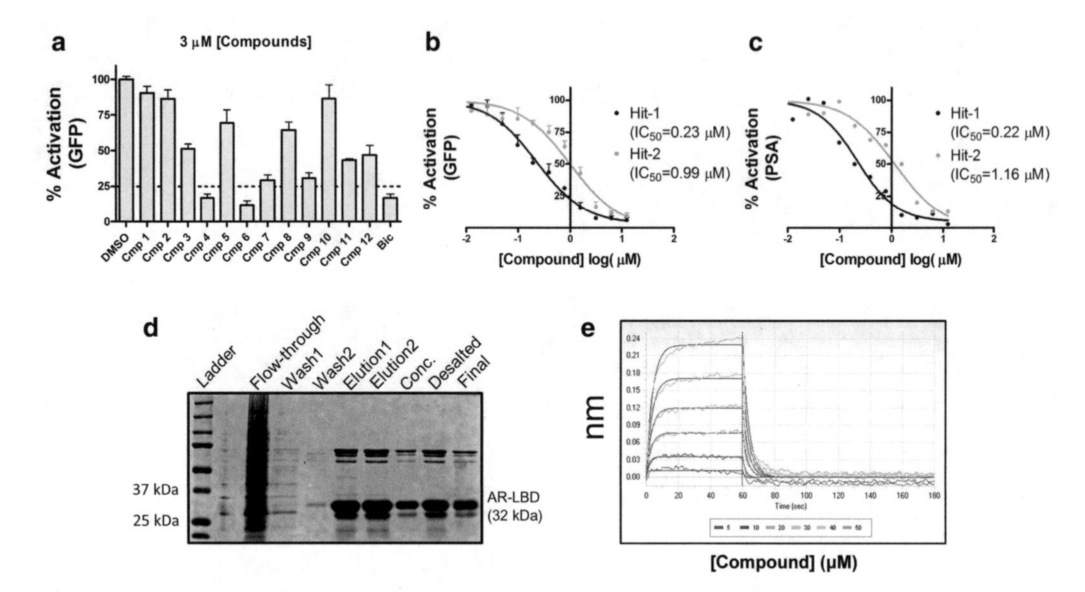

Fig. 5 Cell culture and in vitro experimental validation of inhibitors. (**a**) Single-concentration screening of compounds at 3 μM concentration in LNCaP-eGFP cells. The transcriptional activity of the endogenously expressed AR in LNCaP cells is used to determine the amount of eGFP expression after compound treatment. Error bars represent the mean and standard deviation from six independent replicates. The dashed line represents the arbitrary threshold of inhibition to select hit compounds. (**b**) Dose-dependent inhibition of the AR (LNCaP-eGFP) by hit compounds at the indicated (*X*-axis) concentration. In this example, Hit-1 = Cmp4 and Hit-2 = Cmp6 from the single-concentration screening in *A. Error bars* represent the mean and standard deviation from three replicates. (**c**) Same as *B* but the amount of secreted PSA is determined from the culture media. (**d**) 12 % SDS-PAGE gel of the metal affinity purification of the AR-LBD with N-terminal biotinylation sequence. (**e**) Example BLI data for dose-dependent binding of a hit compound to the purified, biotinylated AR-LBD linked to streptavidin sensors. *Numbers* below the graph specify the compound concentration for each experiment

interface provides several ways of opening the .mdb file. Choose *MOE > File | New | Database* in MOE. Name the file as ZINC lead-like.mdb. Then, import SD file into the mdb file using *MOE > Database Viewer | File | Import.*

3. It may take several hours to import 4 million compounds into the "ZINC lead-like.mdb" (*see* **Note 5**).
4. In order to ensure that each compound is in a form suitable for subsequent modeling steps such as protein–ligand docking, it is necessary to apply a set of cleaning rules such as removing extraneous salts, removal of minor components, or adjusting protonation states. To do so, go to *MOE > Database Viewer | Compute | Molecule | Wash* and click *OK* (use default settings for all adjustable parameters) (Fig. 2).
5. The *MOE > Database Viewer | Compute | Energy Minimize* command opens the Database Minimize panel (Fig. 2).
6. Use default settings to minimize the energy of the compounds. Adjust the force field to MMFF94X. The purpose of the

minimization process is to build 3D structures and/or minimize the energy of each compound in the database. To calculate energy, each molecule is copied to the MOE Window in turn and minimized. The whole process may take several hours to 3 days.

7. Once the minimized process is completed, export the "ZINC lead-like.mdb" to "ZINC lead-like-minimized.sdf" using *MOE > Database Viewer | File | Save as*. Select *output* to *SD file*.

3.4 Molecular Docking Using Glide SP

1. Molecular docking is one of the major tools in computer-assisted drug design. The goal of protein–ligand docking is to predict the predominant binding mode(s) of a compound/ligand to a protein of known three-dimensional structure.
2. Large-scale docking is performed using a computer cluster consisting of 2128 CPU cores with a total of 400 TB of central hard drives (only 96 CPU cores are used in this study).
3. Please note that Glide SP does not consider crystal ligands (**17 W** in this case) while performing docking studies. So no need to delete it from the receptor.
4. Click the *Clear* workspace toolbar button in Maestro suite.
5. Go to *Maestro> Tasks | Docking | Ligand docking*. The Ligand Docking panel opens with the *Settings* tab displayed. In the *Receptor grid* section, click the *Browse* button and navigate to the directory, choose "4HLW-BF3-grid.zip," and click (Fig. 3).
6. In the *Docking* section, ensure that the Precision option is SP (standard precision).
7. Under *Options*, ensure that Dock flexibly and Sample Ring Conformations are selected. Additionally, penalize non-planar conformation is chosen from the Amide bonds option menu.
8. In the *Ligands* tab, ensure that *File* is selected. *Browse* to select "ZINC lead-like-minimized.sdf."
9. Ensure that the selected *Range* is from 1 to End (the default).
10. Use default settings for all other adjustable parameters in the remaining three tabs, *Core*, *Constraints*, and *Similarity*.
11. The last tab *Output* allows the specification of the type of file to create for the output ligand poses and to determine how many poses to write, per ligand and per docking job. Select *Write pose viewer file (includes receptor)* and *Write structures in Maestro format*.
12. Ensure that the value of the *Write out at most poses per ligand* text box is 1 (the default). Use default settings for all other adjustable parameters.
13. Change the *job name* to ZINC-Glide docking. Specify the number of clusters (96 in this case) under *Ligand docking-job settings* and click *Save and Run*.

14. The estimated time to complete the docking of four million compounds is 2–3 weeks.
15. Once the docking run is completed, results are written to "ZINC-Glide docking_pv.maegz."
16. Import the results from "ZINC-Glide docking_pv.maegz" using *Maestro* > *Project* | *Import structures.*
17. Approximately two million compounds having glide score below –5.0 are selected and exported as "ZINC-Glide docking_pv.sdf" using *Maestro* > *Project* | *Export structures.*
18. Please note that the cutoff value may vary according to shape and depth of the active site on the receptor. In this case, by applying –5.0 (a reasonable cutoff), we obtained a handful number of compounds for further analysis. In case of a deeply buried pocket (unlike BF3 site), cutoff can be increased to –6.0 or –7.0.

3.5 Molecular Docking Using eHiTS

1. Compounds from "ZINC-Glide docking_pv.sdf" are re-docked into the same binding cavity (BF3) using the electronic high-throughput screening (eHiTS) docking module.
2. *In-house* scripts are used to perform eHiTS docking.
3. The estimated time to complete the docking of two million molecules is 1–2 weeks. A total of 500,000 structures which received eHiTS docking score below –3.0 threshold are selected and saved as "ZINC-ehits docking.sdf."
4. Please note that the cutoff value may vary according to shape and depth of the active site on the receptor. In this case, by applying –3.0 (a reasonable cutoff), we obtained a handful number of compounds for further analysis. In case of a deeply buried pocket (unlike BF3 site), cutoff can be increased to –5.0 or –6.0.

3.6 RMSD Calculation

1. In order to identify the most probable docking conformations, root mean square deviation (R.M.S.D.) between conformations generated by Glide SP and eHiTS program is calculated for each molecule.
2. MOE SVL script *mol_rmsd.svl* is downloaded from SVL Exchange (http://svl.chemcomp.com/).
3. Convert both "ZINC-Glide docking_pv. sdf" and "ZINC-ehits docking.sdf" into respective mdb files in MOE as described in Subheading 3.3, **step 2**.
4. Load *mol_rmsd.svl* using *MOE* > *File* | *Open*. This command opens the *Dock RMSD Calculator* panel. Specify mdb files converted in **step 3**.
5. A new column is created in the mdb file containing RMSD values. Compounds that obtained RMSD value less than 2 Å are selected for further consensus scoring.
6. Save this file as "4HLW-dock-rmsd.mdb."

3.7 Consensus Scoring

1. The free energy of binding represents a pivotal criterion for compound selection in the drug-discovery process. The free energies of binding predicted by scoring functions can be used to make assumptions about activity, selectivity, and toxicity of drug candidates. In this case, in order to avoid any bias from Glide and eHiTS scoring functions, the generated docking poses are further evaluated using two additional scoring metrics.
2. *pKi prediction*: pKi for 500,000 docked compounds was predicted using MOE SVL script *scoring.svl* implemented in MOE 2012. This script is downloaded from the SVL Exchange. pKi prediction improves accuracy of the prediction of energies of hydrogen bonds and hydrophobic interactions between the atoms of ligand and binding pockets.
3. *Ligand Explorer (LigX) score*: LigX score accounts for the flexibility of AR BF3 pocket and docked conformations of the compounds. LigX score for 500,000 docked compounds is predicted using LigX module-implemented MOE 2012. This function is executed by *MOE > Compute | Simulations | Dock* to compute LigX score. Default settings are used.
4. *Voting*: On the basis of five scores (Glide SP, eHiTS, pKi, LigX, and RMSD) obtained from the above procedures, each molecule is given a vote of 1 or 0 for every "top 10% appearance." First, compounds are ranked according to Glide SP score. A vote of 1 is given to compounds that fall in 10% of the database. A vote of 0 is given to rest of the compounds. This approach was repeated using eHiTS, pKi, LigX, and RMSD scores.
5. Votes from Glide SP, eHiTS, pKi, LigX, and RMSD scores are calculated for each compound. The final cumulative vote (with the maximum possible value of 5) is then used to rank all the docked compounds (Fig. 4).
6. On the basis of the cumulative count, most highly voted molecules (approximately 5000) are selected, and their docking poses are subjected to visual inspection. The aim of the final visual inspection is to select compounds that have good fitness of binding in the BF3 pocket and to eliminate compounds that have toxic or reactive functional groups such as alkyl halides and aldehydes.
7. Based on the availability of compounds for purchase, we selected 100 compounds to be tested using in vitro assays.

3.8 Screening for Transcriptional Inhibition of the AR in Cell Culture

1. LNCaP-eGFP cells are serum deprived in RPMI + CSS for 5 days prior to experiment and maintained in 10 cm cell culture plates (37 °C, 5% CO_2). Handling of cells takes place in a sterile bio-safety cabinet with upward airflow.
2. Cells are prepared for seeding by aspirating the media, adding 3.5 mL pre-warmed trypsin for 5 min, and neutralizing the

proteolysis with 6.5 mL of RPMI + CSS. The cell suspension is transferred to a 50 mL tube and centrifuged at 1000 × *g* (4 min).

3. The cell pellet is resuspended in 4 mL of RPMI + CSS, of which 10 μL is loaded into each well of a dual-chamber cell counting slide for analysis on the TC-10 cell counting instrument.
4. Cells are diluted to 200,000 cells/mL (RPMI + CSS) in preparation for seeding on a black 96-well plate with clear glass bottom (termed the **cell plate**) according to Table 1.
   (a) 11 mL of cell suspension is supplemented with 0.2 nM R1881 and transferred to a 25 mL reservoir. 100 μL (20,000 cells) of suspension is pipetted (multichannel pipet) to wells A1–F12, G4–G6, and H4–H6.
   (b) 1 mL of cell suspension is supplemented with 0.2 % ethanol and pipetted into (100 μL) wells G1–G3 and H1–H3.
   (c) 1 mL of cell suspension is supplemented with 2 nM R1881 and pipetted (100 μL) into wells G10–G12 and H10–H12.
   (d) 1 mL of cell suspension is supplemented with 100 μM final bicalutamide (diluted from 100 mM stock) and 0.2 nM R1881. 100 μL is transferred to wells G7–G9 and H7–H9.
5. A clear 96-well plate containing compound dilutions (termed the **compound plate**) is assembled according to the layout in Table 1. Each well contains 110 μL of RPMI + CSS + 1 % DMSO supplemented with 6 μM of candidate compound (wells A1–F12) or without compound (wells G1–H12). Each plate can screen 12 different candidate compounds with 6 replicates, or 3 replicates of 24 compounds (*see* **Note 6**).
6. Immediately after seeding, 100 μL of each well of the **compound plate** is transferred to the corresponding wells of the **cell plate** that was seeded in **step 4**.
7. The **cell plate** is gently shaken by hand and then incubated for 3 days at 37 °C.
8. Fluorescence from expressed GFP is read on a TECAN m200Pro plate reader running i-control 1.6 for infinity 500 software. The following settings are used: excitation wavelength = 485 nm; emission wavelength = 535 nm; 20 flashes; and multiple reads per well = square 2 × 2.
9. The data obtained from a typical experiment is processed as follows and as shown in Fig. 5a:
   (a) The GFP signal from each well is processed by subtracting the average fluorescence obtained from the control wells lacking R1881 (wells G1–G3, H1–H3).
   (b) The level of GFP expression without inhibitors is calculated from the average fluorescence from wells G4–G6 and

Table 1

Typical experimental setup for single-concentration screening of anti-AR compounds in eGFP-LNCaP cells. Final assay concentrations are given

	1	2	3	4	5	6	7	8	9	10	11	12
A	3 μM	3 μM	3 μM	3 μM	3 μM	3 μM	3 μM	3 μM	3 μM	3 μM	3 μM	3 μM
B	Cmp 1	Cmp 2	Cmp 3	Cmp 4	Cmp 5	Cmp 6	Cmp 7	Cmp 8	Cmp 9	Cmp 10	Cmp 11	Cmp 12
C	0.1 nM	0.1 nM	0.1 nM	0.1 nM	0.1 nM	0.1 nM	0.1 nM	0.1 nM	0.1 nM	0.1 nM R1881	0.1 nM R1881	0.1 nM
D	R1881	R1881	R1881	R1881	R1881	R1881	R1881	R1881	R1881			R1881
E												
F												
G	0 nM R1881			0.1 nM R1881			0.1 nM R1881			1 nM R1881		
H	(or Ethanol 100%)						+50 μM Bicalutamide					

H4–H6. This average is considered as 100% of AR transcriptional activity.

(c) Positive control experiments with bicalutamide (wells G7–G9, H7–H9) will demonstrate whether a known inhibitor of the AR yields the expected effect on GFP transcription.

(d) R1881 control at 1 nM (wells G10–H12, H10–H12) will confirm that an increased dose of the androgen increases GFP transcription and fluorescence.

3.9 Dose-Dependent Inhibition of Hit Compounds

1. Hit compounds from Subheading 3.8 inhibiting the AR transcriptional activity by 75% or greater are selected for further testing (*see* **Note 7**).
2. LNCaP-eGFP cells are maintained and seeded at 20,000 cells/well into an entire black 96-well plate with clear glass bottom (**cell plate**) according to Subheading 3.8, **steps 1–4**.
3. Compounds are initially diluted to 25 μM in a 24-well plate by mixing 0.5 μL of the 50 mM hit compound stock with 1 mL of RPMI + CSS and adjusted to 1% DMSO.
4. 230 μL of the 25 μM hit compound dilutions are transferred in triplicate to the first column of the **compound plate** (clear 96-well), in which twofold serial dilutions are performed in 110 μL of RPMI + CSS and 1% DMSO, as shown in Table 2. Rows G and H of the compound plate are filled with 110 μL of RPMI + CSS and 1% DMSO. Two different hit compounds can be tested in a plate (*see* **Note 8**).
6. Cell plate is gently shaken by hand and then incubated for 3 days at 37 °C.
7. GFP fluorescence is read and interpreted according to Subheading 3.8, **step 9**.
8. Following GFP analysis, 80 μL of media from a well at each compound concentration of the **cell plate** (and from negative controls) is mixed with an equal volume of PBS in Cobas 2 mL measurement vials.
9. Determination of secreted PSA levels at each compound concentration is performed on the Cobas e411 analyzer. 100% of secreted PSA is calculated from the average measurements from wells G4–G6 and H4–H6.
10. The GFP fluorescence and PSA inhibition are plotted in Fig. 5b, c, respectively, and fitted to a sigmoidal (variable slope) dose-dependent curve according to the following equation: $Y = \text{Lower} + (\text{Upper} - \text{Lower}) / \left(1 + 10^{((\text{LogIC}_{50} - X) \times \text{HillSlope})}\right)$, where X is the logarithm of compound concentration, Y is the normalized fluorescence or PSA levels (%Activation or %PSA), and Y starts at Lower and goes to Upper with a sigmoid shape. IC_{50} values are determined after curve fitting.

Table 2
Typical experimental setup for dose-dependent screening of anti-AR compounds. Final assay concentrations are given

	1	2	3	4	5	6	7	8	9	10	11	12
A B C	12.5 μM Cmp 1	6.25 μM Cmp 1	3.13 μM Cmp 1	1.56 μM Cmp 1	0.78 μM Cmp 1	0.39 μM Cmp 1	0.20 μM Cmp 1	0.09 μM Cmp 1	0.05 μM Cmp 1	0.02 μM Cmp 1	0.01 μM Cmp 1	5 nM Cmp 1
D E F	12.5 μM Cmp 2	6.25 μM Cmp 2	3.13 μM Cmp 2	1.56 μM Cmp 2	0.78 μM Cmp 2	0.39 μM Cmp 2	0.20 μM Cmp 2	0.09 μM Cmp 2	0.05 μM Cmp 2	0.02 μM Cmp 2	0.01 μM Cmp 2	5 nM Cmp 2
G H	0 nM R1881 (or ethanol 100%)			0.1 nM R1881			0.1 nM R1881 +50 μM Bicalutamide			1 nM R1881		

3.10 Purification of the AR-LBD

1. 4 L of LB, supplemented with 100 μg/mL ampicillin and 35 μg/mL chloramphenicol, is inoculated with 40 mL of an overnight culture of *E. coli* BL21(DE3) co-transformed with plasmid Pan4-AR-LBD and pBirAcm biotin ligase (*see* Subheading 2.4, **item 1**).
2. The cells are shaken at 37 °C until an OD_{600} nm of 0.6.
3. The cells are placed in a shaker at 16 °C and cooled for 30 min. DHT, IPTG, and biotin are added to the culture to a final concentration of 20 μM, 100 μM, and 125 μM, respectively.
4. The cells are incubated overnight at 16 °C (*see* **Note 9**). All subsequent steps are performed at 4 °C.
5. Cells are pelleted by centrifugation at 5000 × *g* (10 min), followed by resuspension in 50 mL lysis buffer in which two crushed protease inhibitor tablets are dissolved.
6. The cells are broken by sonication with ten pulses of 30 s each, with 30-s cooling on ice between pulses. The unbroken cells are removed by centrifugation at 18,000 × *g* (30 min) (*see* **Note 10**).
7. 2 mL of Ni–NTA agarose beads are added to the lysates and gently rotated for 1 h.
8. The lysate/Ni-NTA mixture is poured into a 20 mL poly-prep gravity chromatography column. The flow-through is kept for later analysis by SDS-PAGE.
9. The beads are washed three times with 5 mL wash buffer (containing 20 mM imidazole). Washed flow-through is saved for later analysis by SDS-PAGE (**step 13**).
10. Purified biotinylated AR-LBD is eluted from the columns with 4 mL of elution buffer (containing 250 mM imidazole) and collected in 2 mL fractions. Approximately 50 μL of each elution is saved for analysis by SDS-PAGE (**step 13**).
11. Elution fractions are pooled and mixed with dilution buffer until the final imidazole concentration is 100 mM or lower. The solution is transferred to a chilled 15 mL, 10 kDa Amicon concentrator and centrifuged for ~20 min at 4000 × *g*, 4 °C, in a swinging-bucket centrifuge (*see* **Note 11**). Save 50 μl for later analysis in **step 13**.
12. The Zeba desalting column (Subheading 2.4, **item 16**) is equilibrated with 3 × 5 mL of dilution buffer by centrifugation at 2000 × *g* (2 min). The concentrated AR-LBD is then desalted on the equilibrated column by centrifugation at 2000 × *g* (2 min). The AR-LBD is spun a final time in the 10 kDa centricon concentrator at 4000 × *g* (10 min). At each step, 50 μL of sample can be saved for analysis in **step 13**.
13. 50 μL of flow-through, washes, elutions, and concentrated/desalted AR-LBD are mixed with 10 μL of 5× sample buffer

and separated on a 12% SDS-PAGE gel (Fig. 5d). Instructions for mounting and running an SDS-PAGE mini gel (Bio-Rad) are given at http://www.bio-rad.com/LifeScience/pdf/Bulletin_4006193A.pdf.

14. The final biotinylated AR-LBD is aliquoted, flash-frozen in liquid nitrogen, and stored at −80 °C for use in BLI experiments (*see* **Note 12**).

3.11 BLI Analysis of Compound Binding to the AR-LBD

1. The biotinylated AR-LBD is diluted to 0.1 g/L concentration in BLI buffer.
2. Sensor needles are placed into columns 1 and 2 of the FortéBIO green tray. The lid may be used at this step to protect the sensors.
3. The appropriate buffers and AR-LBD protein are added to a black 96-well plate (termed the **sensor plate**) according to Table 3. All wells contain 200 μL of volume: Column (C) 1–2, BLI buffer; C3, 0.1 g/L AR-LBD; C4, BLI buffer supplemented with 1000× diluted biocytin; C5-6, 1000× diluted biocytin in SuperBlock Blocking Buffer in TBS; C7–8, BLI buffer; and C9–10, BLI buffer supplemented with 1% DMSO.
4. The green rack is placed on top of the black 96-well plate, allowing the sensors to dip into the contents of C1–2. Incubate for 20 min at room temperature (r.t.)
5. Sensors are moved to corresponding positions from C1 to C3 and C2 to C4. This step will load the purified AR-LBD protein onto one set of sensors (C3) whereas the other sensors (C4) will be blocked with biocytin and will serve as a reference. Incubate overnight at 4 °C.
6. Sensors are moved to corresponding position from C3 to C5 and C4 to C6. This step blocks any remaining free streptavidin-binding sites. Incubate for 1 h at r.t.

Table 3
BLI sensor plate experimental setup

	1	2	3	4	5	6	7	8	9	10
A	BLI buffer		AR-LBD	Biocytin	Biocytin in SuperBlock TBS		BLI buffer		BLI buffer + 1% DMSO	
B										
C										
D										
E										
F										
G										
H										

Table 4
BLI sample plate experimental setup

	1	2	3	4	5	6	7	8	9	10	11	12	
A	BLI buffer + 1 % DMSO						5 μM	10 μM	20 μM	30 μM	40 μM	50 μM	Cmp 1
B							5 μM	10 μM	20 μM	30 μM	40 μM	50 μM	Cmp 2
C							5 μM	10 μM	20 μM	30 μM	40 μM	50 μM	Cmp 3
D							5 μM	10 μM	20 μM	30 μM	40 μM	50 μM	Cmp 4
E							5 μM	10 μM	20 μM	30 μM	40 μM	50 μM	Cmp 5
F							5 μM	10 μM	20 μM	30 μM	40 μM	50 μM	Cmp 6
G							5 μM	10 μM	20 μM	30 μM	40 μM	50 μM	Positive control
H							BLI buffer + 1 % DMSO						DMSO 1 %

7. Sensors are moved to corresponding position from C5 to C7 and C6 to C8 to wash away excess biocytin. Incubate for 2 min at r.t.
8. Sensors are moved to corresponding position from C7 to C9 and C8 to C10 to equilibrate in BLI buffer containing 1 % DMSO.
9. The entire assembly, without the lid, is clipped into place on the left slot in the Octet Red instrument.
10. A black 96-well plate bearing compounds, termed the **sample plate**, is assembled according to Table 4. All wells contain 200 μL of volume: C1–C6, BLI buffer + 1 % DMSO; C7–C12, 5, 10, 20, 30, 40, and 50 μM of compound, respectively. Row H contains only BLI buffer + 1 % DMSO. Each row can hold a different compound allowing for 6 assays to be run simultaneously together with a positive control in row 7 (*see* **Note 13**). The **sample plate** is placed in the right slot of the Octet Red instrument.

3.11.1 Acquiring Data with the BLI Software

1. The FortéBIO Data Acquisition Windows™ software is opened and "New Kinetics Experiment" is selected (*see* **Note 14**).
2. Under the "Plate definition" tab, the sample ID, compound identity, and concentration are specified for the **sample plate**. Wells without compounds are set as "buffer wells."
3. Definitions are created under the "Assay definition" tab to control the sequential steps to run the experiment:

   Baseline 1, Time = 180 s, Shake = 1000.

   Baseline 2: Time = 60 s, Shake = 1000.

   Association, Time = 60 s, Shake = 1000.

   Dissociation, Time = 120 s, Shake = 1000.
4. Once definitions are created, the actual sequence of events can now be programmed into the software:

(a) Sensors are first moved to C1 of the **sample plate** to equilibrate in buffer and acquire "Baseline 1." Baseline 1 is only done once at the beginning of the experiment.

(b) "Baseline 2" is acquired and is repeated before any association measurement.

(c) Sensors are moved to C7 for "Association" of compound at 5 μM concentration.

(d) Sensors are moved back to C1 for "Dissociation" of compound.

(e) Steps b-d are repeated using the next columns in the series. For example, the measurement of the second compound concentration in the series will acquire "Baseline 2" in C2, "Acquisition" in C8, and "Dissociation" in C2.

(f) Once the entire sequence is inputted, the "replicate" checkbox is marked and the "Add new assay" option is selected to repeat the same sequence with the second set of biocytin-blocked sensors.

5. Under the "Sensor Assignment" tab, exclude C1–C8 of the **sensor plate** by selecting these wells and identifying them as "Blocked." Manually add the information for C9 as protein AR-LBD and C10 as Biocytin.
6. Under the "Review Experiment" tab, review the sequence of events, compound identity, and concentrations.
7. Under the "Run Experiment: tab," start the experiment with "Delay eq. start = 300 s" and temperature set at 30 °C. The folder into which data will be saved is also selected during this step.
8. Several methods exist to analyze the data once acquired of which we describe a typical scenario. The "ForteBIO Data Analysis Software" is opened and the acquired data is opened using the "Load Folder" option.
9. Data processing involves indicating to the software which wells in the sensor and compound plates represent samples or reference controls:

(a) In the **sensor plate** map, biocytin control sensors are selected (C10). Right click, select "Change Sensor Type," and choose "Reference." This will subtract the BLI signal of the biocytin-blocked sensor from the AR-LBD-bearing sensor.

(b) In the **sample plate** map, C1–6 is identified as "Buffer" and H7–H12 is identified as "Reference."

(c) On the left panel of the software, "Subtraction – Double reference" is checked.

(d) "Align Y axis" is checked where the corresponding time frame is set to consider only the last 5 s of acquisition (55–59.9 s).

(e) "Inter-step correction – Align to dissociation" is checked.

(f) "Process Data" is clicked.

10. Data analysis involves grouping the processed data:

(a) Group graphs by "Sample ID" and second group by "K_D (M)."

(b) Change the legend to "Concentration (μM)."

(c) Under Data Options, check "Use Included Traces Only," "Show Curve Fits," and "Display Traces in Table Color."

11. A typical graph of processed and grouped data for showing a dose-dependent binding of a single compound to the AR-LBD is shown in Fig. 5e.

4 Notes

1. If the LNCaP-eGFP cell line is not available, a reporter plasmid could be transiently transfected into LNCaP cells, such as a luciferase or GFP construct under the control of probasin-based or PSA-based AR-regulated promoter. For studies with other steroid/nuclear receptors, appropriate reporter constructs should be chosen.
2. Purification attempts of the LBD in the presence of DHT gave better yield and more stable protein than in the presence of R1881.
3. To study the direct binding between the isolated AR-LBD domain and the BF3 inhibitors, we employ the BLI technique available in our laboratory. Alternatively, other techniques can be used for the same purpose such as isothermal titration calorimetry (ITC) or surface plasmon resonance (SPR).
4. This process can be followed to prepare crystal structures of other nuclear receptors.
5. Importing ZINC lead-like.sdf (~4 million compounds) into MOE might overrun your computer capabilities. To avoid it, the SDF file can be split into several smaller files. Importing a small SDF file can be more convenient.
6. The screening compound concentration can be modified to allow for more (lesser concentration) or less (higher concentration) stringent criteria to identify hit compounds.
7. The cutoff threshold is arbitrary but lowering beyond 75 % can lead to many false positives.
8. The dilution range of compound concentrations can be modified according to the current compound activity and expected IC_{50}. The experiment can be repeated with enzalutamide,

bicalutamide, or other antiandrogen with known IC_{50} values to serve as a positive control.

9. Different times of induction are possible. The minimum induction time was found to be 8 h at 16 °C. The lower temperature helps the AR-LBD to remain stable as it is produced inside the bacteria.
10. If the cell lysate is too viscous or if particulate matter is still present after centrifugation, the lysate can be passed to a syringe filter with 0.22 μM pore size. Alternatively, the lysate can be sonicated for additional pulses or centrifuged for a longer period of time.
11. The AR-LBD is prone to precipitation which may clog the Amicon device. Periodically, the concentration process should be monitored to make sure that the protein does not fall out of solution. Shorter cycles of centrifugation (5 min/cycle) and mixing the protein by pipetting between the cycles will help to monitor the behavior of the protein and halt any emerging precipitation.
12. This purification protocol results in sufficient yield of protein in order to screen a large number of potential inhibitors using the BLI technique. In case of comparison between different proteins (e.g., wild type versus mutant), it is preferable to include a size-exclusion chromatography step in the purification protocol to obtain pure, monodisperse protein.
13. Positive BLI control for AR-LBD binding is typically a known antiandrogen such as bicalutamide or an AF2-binding peptide such as a peptide from SRC (steroid receptor co-activator).
14. The full manual for the Octet Red instrument and software can be found at http://www.fortebio.com/documents/Octet_Users_Guide_PN_41-0000rev290806.pdf.

Acknowledgements

This work was funded by Prostate Cancer Canada with generous support from Canada Safeway—Grant SP2013-02. This research was also supported by the Department of Defense (Prostate Cancer Res. Program) under award number (W81XWH-12-1-0401). Views and opinions of, and endorsements by, the authors do not reflect those of the US Army or the Department of Defense. The authors acknowledge the financial support from Canadian Institutes of Health Research and Canadian Cancer Society Research Institute grant F12-03271. R. Munuganti would like to thank Prostate Cancer Foundation-British Columbia (PCFBC), Canada, for providing graduate fellowship. K. Dalal is supported by CIHR and MSFHR postdoctoral fellowships.

References

1. Green SM, Mostaghel EA, Nelson PS (2012) Androgen action and metabolism in prostate cancer. Mol Cell Endocrinol 360:3–13
2. Lallous N, Dalal K, Cherkasov A, Rennie PS (2013) Targeting alternative sites on the androgen receptor to treat castration-resistant prostate cancer. Int J Mol Sci 14:12496–12519
3. Jenster G, van der Korput HA, van Vroonhoven C, van der Kwast TH, Trapman J, Brinkmann AO (1991) Domains of the human androgen receptor involved in steroid binding, transcriptional activation, and subcellular localization. Mol Endocrinol 5:1396–1404
4. Cutress ML, Whitaker HC, Mills IG, Stewart M, Neal DE (2008) Structural basis for the nuclear import of the human androgen receptor. J Cell Sci 121:957–968
5. Lamont KR, Tindall DJ (2010) Androgen regulation of gene expression. Adv Cancer Res 107:137–162
6. Saad F, Miller K (2013) Treatment options in castration-resistant prostate cancer: current therapies and emerging docetaxel-based regimens. Urol Oncol 32(2):70–79
7. Bohl CE, Gao W, Miller DD, Bell CE, Dalton JT (2005) Structural basis for antagonism and resistance of bicalutamide in prostate cancer. Proc Natl Acad Sci U S A 102:6201–6206
8. Salvati ME, Balog A, Wei DD, Pickering D, Attar RM, Geng J, Rizzo CA, Hunt JT, Gottardis MM, Weinmann R, Martinez R (2005) Identification of a novel class of androgen receptor antagonists based on the bicyclic-1H-isoindole-1,3(2H)-dione nucleus. Bioorg Med Chem Lett 15:389–393
9. Duke CB, Jones A, Bohl CE, Dalton JT, Miller DD (2011) Unexpected binding orientation of bulky-B-ring anti-androgens and implications for future drug targets. J Med Chem 54: 3973–3976
10. He B, Gampe RT Jr, Kole AJ, Hnat AT, Stanley TB, An G, Stewart EL, Kalman RI, Minges JT, Wilson EM (2004) Structural basis for androgen receptor interdomain and coactivator interactions suggests a transition in nuclear receptor activation function dominance. Mol Cell 16:425–438
11. Estebanez-Perpina E, Arnold LA, Nguyen P, Rodrigues ED, Mar E, Bateman R, Pallai P, Shokat KM, Baxter JD, Guy RK, Webb P, Fletterick RJ (2007) A surface on the androgen receptor that allosterically regulates coactivator binding. Proc Natl Acad Sci U S A 104: 16074–16079
12. Lack NA, Axerio-Cilies P, Tavassoli P, Han FQ, Chan KH, Feau C, LeBlanc E, Guns ET, Guy RK, Rennie PS, Cherkasov A (2011) Targeting the binding function 3 (BF3) site of the human androgen receptor through virtual screening. J Med Chem 54:8563–8573
13. Jehle K, Cato L, Neeb A, Muhle-Goll C, Jung N, Smith EW, Buzon V, Carbo LR, Estebanez-Perpina E, Schmitz K, Fruk L, Luy B, Chen Y, Cox MB, Brase S, Brown M, Cato AC (2014) Coregulator control of androgen receptor action by a novel nuclear receptor-binding motif. J Biol Chem 289(13):8839–51
14. Munuganti RS, Leblanc E, Axerio-Cilies P, Labriere C, Frewin K, Singh K, Hassona MD, Lack NA, Li H, Ban F, Tomlinson Guns E, Young R, Rennie PS, Cherkasov A (2013) Targeting the binding function 3 (BF3) site of the androgen receptor through virtual screening. 2. development of 2-((2-phenoxyethyl) thio)-1H-benzimidazole derivatives. J Med Chem 56:1136–1148
15. Li H, Ren X, Leblanc E, Frewin K, Rennie PS, Cherkasov A (2013) Identification of novel androgen receptor antagonists using structure- and ligand-based methods. J Chem Inf Model 53:123–130
16. Li H, Ban F, Dalal K, LeBlanc E, Frewin K, Ma D, Rennie PS, Cherkasov A (2014) Discovery of small-molecule inhibitors selectively targeting the DNA-binding domain of the human androgen receptor. J Med Chem 57: 6458–6467
17. Dalal K, Roshan-Moniri M, Sharma A, Li H, Ban F, Hessein M, Hsing M, Singh K, LeBlanc E, Dehm S, Guns ET, Cherkasov A, Rennie PS (2014) Selectively targeting the DNA binding domain of the androgen receptor as a prospective therapy for prostate cancer. J Biol Chem. doi:10.1074/jbc.M114.553818

Part III

Nuclear Receptor Signalling

Chapter 5

Use of BRET to Study Protein–Protein Interactions In Vitro and In Vivo

Shalini Dimri, Soumya Basu, and Abhijit De

Abstract

Application of bioluminescence resonance energy transfer (BRET) assay has been of special value in measuring dynamic events such as protein–protein interactions (PPIs) in vitro *or* in vivo. It was only in the late 1990s the BRET assay using RLuc-YFP was introduced for biological research showing its use in determining interaction of two proteins involved in circadian rhythm. Several inherent attributes such as rapid and fairly sensitive ratiometric measurements, assessment of PPI irrespective of protein location in cellular compartment, and cost-effectiveness consenting to high-throughput assay development make BRET a popular genetic reporter-based assay for PPI studies. In BRET-based screening, within a defined proximity range of 10–100 Å, excited state energy of the luminescence molecule can excite the acceptor fluorophore in the form of resonance energy transfer, causing it to emit at its characteristic emission wavelength. Based on this principle, several such donor–acceptor pairs, using the *Renilla* luciferase or its mutants as donor and either GFP2, YFP, mOrange, TagRFP, or TurboFP as acceptor, have been reported for use.

In recent years, BRET-related research has become significantly versatile in the assay format and its applicability by adopting the assay on multiple detection devices such as small-animal optical imaging platform or bioluminescence microscope. Beyond the scope of quantitative measurement of PPIs and protein dimerization, molecular optical imaging applications based on BRET assays have broadened its scope for screening of pharmacological compounds by unifying in vitro, *live cell, and* in vivo animal/plant measurement all on one platform. Taking examples from the literature, this chapter contributes to in-depth methodological details on how to perform in vitro and in vivo BRET experiments, and illustrates its advantages as a single-format assay.

Key words Bioluminescence resonance energy transfer, Protein–protein interactions, Cell-based assay, Luciferase, Fluorescent proteins, Optical imaging

1 Introduction

Protein–protein interactions form the key molecular process in a biological system and drive almost all the cellular and molecular functions like cell division, cell signaling, immune responses, and response to environmental stimuli. For a better understanding of how cellular functions are regulated, noninvasive measurement of protein–protein

Iain J. McEwan (ed.), *The Nuclear Receptor Superfamily: Methods and Protocols*, Methods in Molecular Biology, vol. 1443, DOI 10.1007/978-1-4939-3724-0_5, © Springer Science+Business Media New York 2016

interactions in a live cell environment is important. To do so, a sensitive and real-time method that can qualitatively as well as quantitatively measure dynamic events in an unperturbed condition is of high demand. Conventional methods used so far to study protein–protein interactions, like chromatography, co-immunoprecipitation, tandem affinity purification, phage display, and chemical cross-linking [1], are unable to provide a direct insight into macromolecular interactions in live cells maintaining the spatial-temporal information intact. The shortcomings of the above approaches have in part been overcome by newer reporter gene-based strategies like inducible yeast two-hybrid systems, bimolecular fluorescence complementation (BiFC), fluorescence resonance energy transfer (FRET), and bioluminescence resonance energy transfer (BRET)—which can provide visual perception to what is happening to proteins inside the cell and in their native environment.

BRET is based on the principle of Forster resonance energy transfer in which the transfer of resonance energy from excited bioluminescent molecule (called donor) to a fluorescent molecule (called acceptor) forms the basis of detection. In the presence of an appropriate substrate, the bioluminescent reporter protein conjugated to one of the proteins of interest oxidizes the substrate reaching it to the excited state. The excited substrate then releases energy, which is taken up by the fluorescent molecule conjugated to the second protein of interest, when the proximity range is achieved by their interaction. The excited fluorophore then emits the characteristic light at a longer wavelength. The nonradiative transfer of energy between donor and acceptor can take place only when the two molecules of interest are in close proximity, i.e., 1–10 nm of distance, which is a distance for true protein–protein interactions in the physiological and biological environment [2]. Hence a positive BRET signal is an actual interpretation to a true protein interaction (*see* Fig. 1). At the same time, however, absence of BRET signal does not necessarily mean that the two proteins are not interacting, rather their interaction simply failed to achieve the necessary proximity [3]. To produce an efficient BRET output signal, the selection and design of BRET partners should fulfill the following conditions: (1) the distance between the donor and acceptor molecule should be less than 10 nm; (2) spectral overlap between the donor emission and acceptor excitation peak wavelength; (3) relative orientation of donor and the acceptor molecule, i.e., either N-terminus or C-terminus localization in which the dipoles of donor and acceptor are aligned in a way that there is maximum transfer of resonance energy through nonradiative dipole–dipole coupling; and (4) donor quantum output; the higher the donor quantum output the better will be the nonradiative transfer of energy to acceptor and the minimum will be the energy loss due to decay [2]. Inside the cell there are thousands of transient and nonspecific interactions taking place. To differentiate

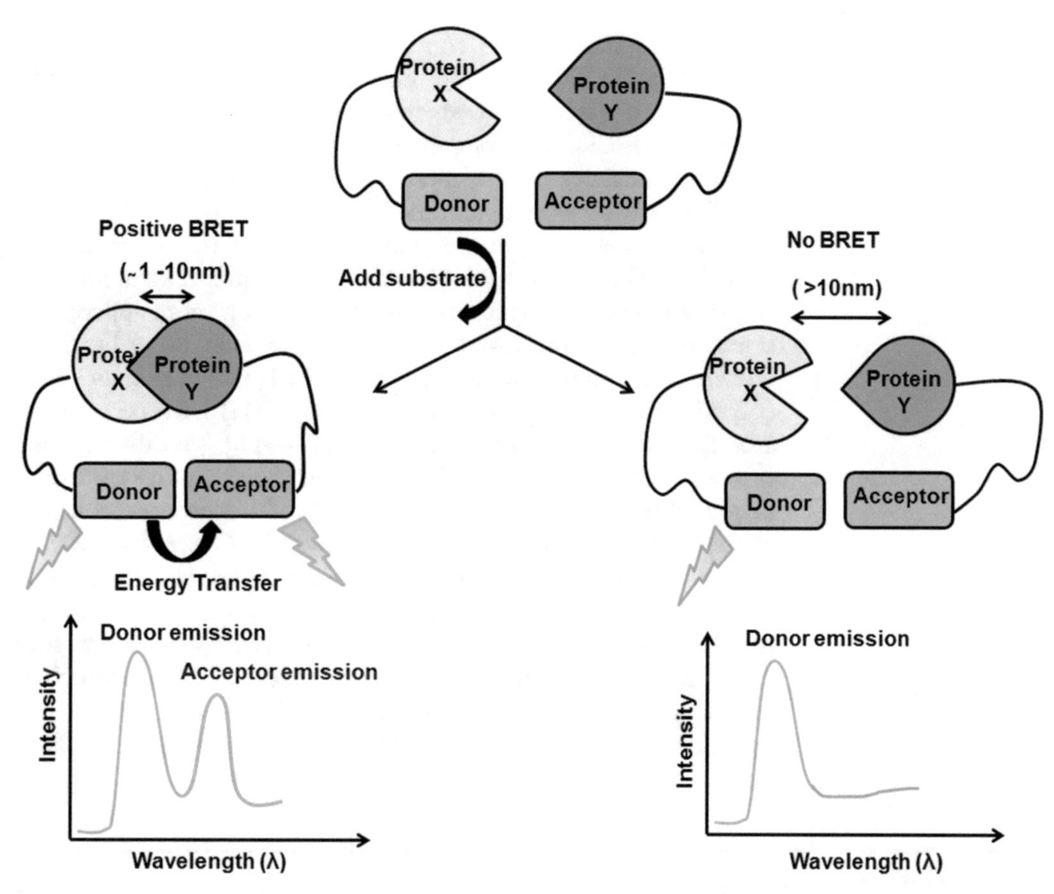

Fig. 1 Diagrammatic representation of a typical BRET assay for determining protein–protein interactions. The protein candidates *X* and *Y* can be tagged with donor and acceptor. If only the two proteins of interest achieve the proximity distance (1–10 nm), BRET occurs in the presence of donor-specific substrate. The transferred resonance energy excites the acceptor fluorophore which then emits at its characteristic wavelength indicating positive interaction of *X* and *Y*. Light signals emitted by both donor and acceptor can be measured by suitable band-pass filters and can be represented ratiometrically as acceptor/donor signal output. If the protein *X* and *Y* fail to achieve the required proximity distance, then only the donor output is obtained

between specific and nonspecific interactions, various formats of BRET assay like donor saturation assay, competition assay, and dilution assay with appropriate positive and negative controls can be performed [4]. These assays not only provide evidence of the specificity of the protein–protein interaction, but can also be extended to study the oligomerization state of receptors [5].

BRET has successfully emerged as a potential, advanced, and noninvasive tool to study a wide variety of assays like protein–protein interactions (e.g., cyanobacterial clock protein-KaiB and light-regulatory basic leucine zipper (bZip) transcription factor-HY5) [6]; oligomerization study of receptors (e.g., GPCRs, receptor tyrosine kinases, and cytokine receptors) [7]; mapping signal transduction pathway; studying protein posttranslational modifications such as ubiquitination [8], sumoylation [9], phosphorylation [10], and acetylation [11]; and monitoring protease activity in live cells [12]. With the advancements made over the years, various modifications in BRET components have enhanced the overall sensitivity and specificity of the method. The short half-life and low stability of Rluc were overcome by introducing a series of point mutations in the enzyme sequence leading to the generation of Rluc 8.6. This mutant version of Rluc has greater stability and red-shifted emission spectrum which makes it a more appropriate donor for animal imaging [13]. Apart from Rluc, other luciferase enzymes such as *Gaussia* luciferase (19.9 KDa), *Vargula* luciferase (62 KDa), and *Oplophorus* luciferase (18 KDa) have been evaluated for use as alternative BRET assays [14–17]. Modified version of coelenterazine substrate like ViviRen™ and EnduRen™ offer brighter and extended signal output [18, 19]. New acceptor fluorophores like TurboRFP635, mOrange, and mCherry have excitation and emission at higher wavelength and hence serve as invaluable elements in expanding BRET application to in vivo animal imaging [20, 21]. Using new version of acceptor, donor, substrate, and instrumentations (BLI microscopy, IVIS), researchers have extended the protein–protein interaction study from in vitro to single cell, and even tissue-scale in vivo imaging both in plants and animals.

2 Materials

2.1 Construction of Fusion Proteins

1. cDNAs for proteins of interest.
2. cDNA for complementary BRET donor and acceptor. Donors for all common BRET systems used so far are *Renilla* luciferase or its mutants: Rluc for $BRET^1$ [22] and $BRET^2$, Rluc2 or Rluc8 for $BRET^3$ [21, 23], and Rluc8.6 for $BRET^8$ [24]. Acceptor for $BRET^1$ is YFP/EYFP, $BRET^2$ is GFP^2, $BRET^3$ is mOrange, and $BRET^8$ is TurboRFP635.
3. For mammalian cell experiments, an expression plasmid such as pcDNA3.1 (+) or similar is required.
4. 1× Passive lysis buffer (Promega, USA).
5. Bradford reagent (BioRad, USA).
6. 1× PBS (pH 7.0).

2.2 Cell Culture

1. Cell culture plates: 6-Well clear cell culture plates.
2. Cell culture plates: 96-Well white cell culture plates.
3. Cell culture plates: 96-Well black cell culture plates.
4. Cell line for transfection: 293-T, HT1080, COS7, Hela, or any specific type.

5. Appropriate media for the cell line: Typically, Dulbecco's modified Eagle's medium (DMEM) containing 0.3 mg/ml glutamine, 100 IU/ml penicillin, 100 mg/ml streptomycin, and 10% fetal bovine serum (FBS) or other specific medium recommended for specific cell type.
6. 0.05% Trypsin–0.53 mM ethylenediamine tetraacetic acid (EDTA).
7. Transfection system or reagent: Effectene (Qiagen, USA), Lipofectamine2000 (Life Technologies, USA), or any other suitable system.

2.3 BRET Assay Ingredients

1. BRET assay buffer: Dulbecco's phosphate-buffered saline (DPBS) containing 0.1 g/l $CaCl_2$, 0.1 g/l $MgCl_2 \cdot 6H_2O$, and 1 g/l d-glucose.
2. Media for BRET measurement: DMEM without phenol red containing 0.3 mg/ml glutamine, 100 IU/ml penicillin, 100 mg/ml streptomycin, 10% FBS, and 25 mM HEPES.
3. Preparation and dilution of luciferase substrate: Coelenterazine h (Promega or Biotium) is reconstituted in methanol at a concentration of 1 mg/ml for $BRET^1$, $BRET^3$, and $BRET^8$ and stored as stock solution in −80 °C freezer. For $BRET^2$, coelenterazine 400a (Molecular Imaging Products Company or Biotium) is reconstituted in anhydrous or absolute ethanol and stored as stock solution (*see* **Note 1**). Just before the experiment, dilute the substrate by adding 10 μg of coelenterazine stock per 100 μl of DPBS. If higher concentration of coelenterazine (80–100 μg) is required, directly dissolve coelenterazine powder in 50% ethanol and 50% PEG mix. EnduRen (Promega) at a stock concentration of 60 mM is reconstituted in cell culture-grade dimethylsulfoxide (Sigma) for eBRET, $BRET^3$, and $BRET^8$ measurement. Extensive vortexing up to 10 min and warming to 37 °C are required during reconstitution of EnduRen. EnduRen stocks can be stored at −20 °C protected from light and moisture.
4. Dilution of luciferase substrate in appropriate assay buffer: Typical assay buffer for coelenterazine h and coelenterazine 400a is d-PBS with $CaCl_2$, $MgCl_2$, and d-glucose (Gibco) and the final concentration of the substrates is 5 μM. Enduren is diluted in a final concentration of 30–60 μM in HEPES-buffered DMEM without phenol red at 37 °C (*see* **Note 2**).
5. Ligand or other modulating reagent: Depending on the interaction being assayed, stock and working solutions are to be made and stored as per the manufacturer's recommendation.
6. Selection of antibiotics such as geniticin (100 mg/ml stock concentration), zeocin (100 mg/ml stock concentration), or puromycin (10 mg/ml stock concentration) depending on the marker present on the vector backbone.

2.4 Animals

For studying protein–protein interactions in live animal subjects, the following considerations should be kept in mind: select animals of same strain, sex (sex has to be determined as per the experimental need), weight, and age group (*see* **Note 3**). For conducting animal experiments generally prior permission is required as per the institutional and national animal ethical guidelines.

2.5 Measurement Equipment

1. Standard cell culture facility including class II biological safety cabinet.
2. 37 °C Incubator with 5 % CO_2.
3. Fluorometer (Fluoroskan Ascent™) or scanning spectrophotometer with 96-well plate capability.
4. Microplate Luminometer like LUMIstar Optima (BMG Labtech, Germany), Mithras LB 940 (Berthold Technologies, Germany), or several other company brands compatible for performing simultaneous dual-channel photon measurement using donor- and acceptor-specific filter sets can be used. Photon measurements can be done using 0.5–5-s integration time per filter (e.g., for $BRET^8$ 540 nm with 10 band-pass as donor and 630 nm with 10 band-pass filter as acceptor when using Enduren or coelenterazine h as substrate). Simultaneous detection increases accuracy and reduces measurement time, and thus would be suitable for high-throughput screening assays.
5. IVIS Lumina, IVIS200, IVIS Spectrum (Perkin Elmer, USA) equipped with 20 nm band-pass spectral filter sets (typically range varies from 450 to 800 nm) or similar other BRET-compatible brands for live cell or in vivo tissue-scale animal imaging.
6. Dual-View microimager (Optical Insights, Tucson, AZ) with modified electron bombardment-CCD camera: It is an emission splitting system that allows user to acquire spectrally distinct but spatially identical images simultaneously. The microimager consists of a dichroic mirror that can split the image into two distinct wavelengths—above and below 505 nm, and the interference filters allow refinement of the distinction [25].
7. Olympus LV200 luminescent microscope (Olympus America, Inc., New York, USA) with respective donor filter and acceptor filters: It has designed optical elements to enhance collection and transmission of light through the specimen. It can co-image phase contrast, transmitted fluorescence, and bright field with luminescence signal and allows detection of localization and co-localization of the luminescence signals with fluorescence probe in tissues or in cells.

3 Methods

3.1 Basic BRET Vector Design and Optimization

1. First select an appropriate donor (luciferase protein) and acceptor (fluorophore) pair with suitable substrate required (*see* Table 1). In many cases N- or C-terminal fusion vectors are available for cloning and expression in mammalian cells. Make sure that for dual selection, different selection markers (e.g., neomycin or zeomycin or puromycin) are inserted in the donor- and acceptor-containing plasmids (*see* Fig. 2).
2. To make fusion constructs PCR amplify the cDNA of target proteins, e.g., X and Y (where X and Y are intended to be interacting partners), donor (e.g., Rluc8.6), and acceptor proteins (e.g., TurboRFP635), flanked by unique restriction sites required for cloning at MCS of expression vector.
3. Insert the cDNA of target protein in frame with cDNA of donor or acceptor protein. If required separate the two proteins using cDNA for a flexible linker. Presence of a linker between

Table 1
Table highlighting the key features of existing and newly developed BRET assays using *Renilla* luciferase. Modified with permission [24]

Assay	Donor	Acceptor	Substrate	Spectral resolution (nm)	Dynamic range	Efficiency
BRET[1]	RLUC *480 nm* (Improved version using RLUC2/ RLUC8)	YFP/EYFP *535 nm*	Clz/Enduren™	55	Small	Moderate
BRET[2]	RLUC *400 nm* (Improved version using RLUC2/ RLUC8)	GFP[2] *515 nm*	Clz400/protected Clz400	115	Very large	Moderate
BRET[3]	RLUC8 *480 nm*	mOrange *564 nm*	Clz/EnduRen™	85	Large	Moderate
BRET[4]	RLUC8 *480 nm*	TagRFP *584 nm*	Clz/EnduRen™	104	Large	High
BRET[5]	RLUC8 *515 nm*	TagRFP *584 nm*	Clz -*v*	70	Moderate	Low
BRET[6]	RLUC8.6 *535 nm*	TagRFP *584 nm*	Clz/EnduRen™	50	Large	High
BRET[7]	RLUC8 *480 nm*	TurboFP *635 nm*	Clz-*v*	155	Small	Low
BRET[8]	RLUC8.6 *535 nm*	TurboFP *635 nm*	Clz/EnduRen™	100	Moderate	Moderate

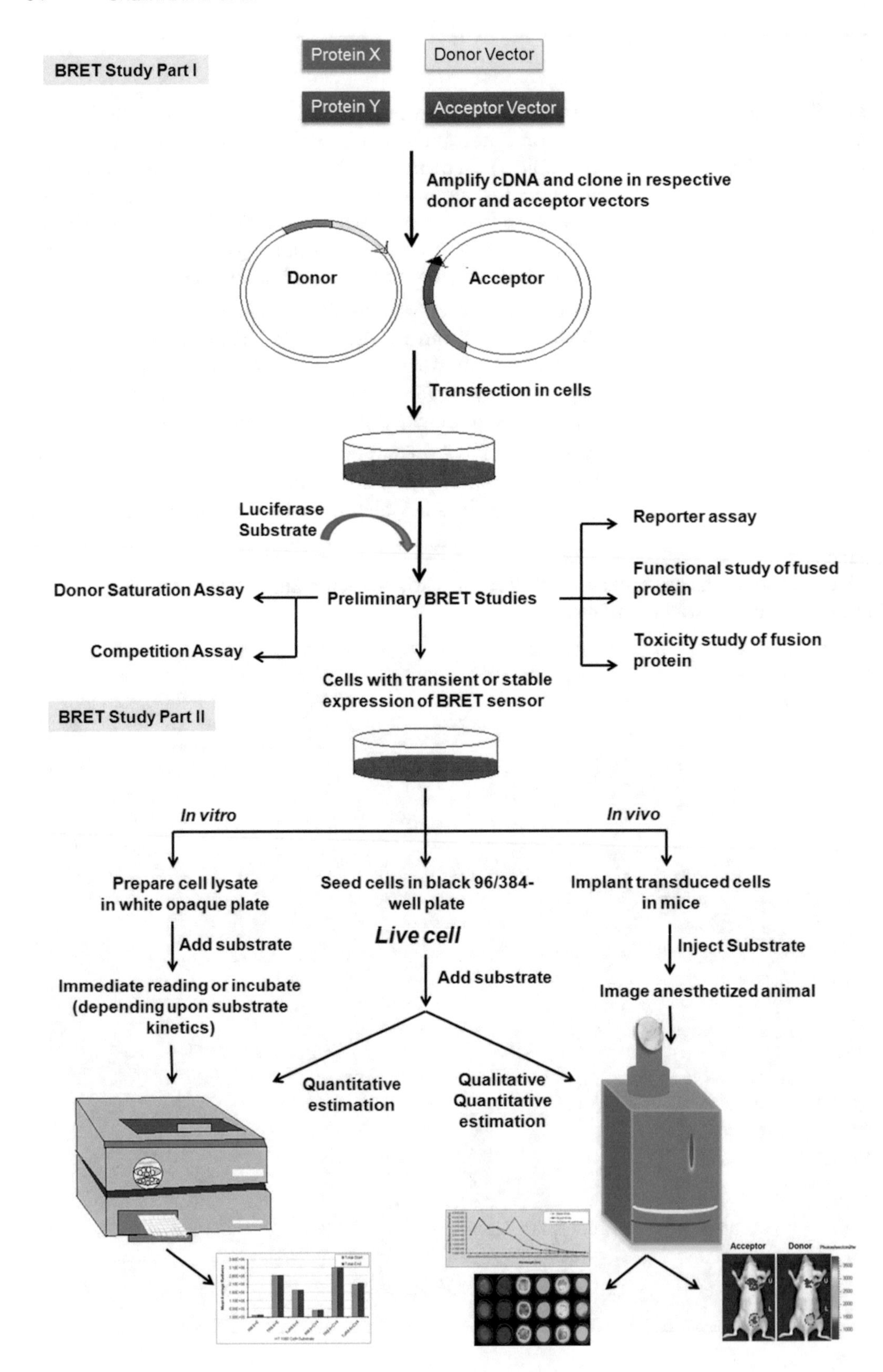
BRET Study Part I
Protein X
Donor Vector
Protein Y
Acceptor Vector
Amplify cDNA and clone in respective donor and acceptor vectors
Donor
Acceptor
Transfection in cells
Luciferase Substrate
Reporter assay
Functional study of fused protein
Donor Saturation Assay
Preliminary BRET Studies
Competition Assay
Toxicity study of fusion protein
Cells with transient or stable expression of BRET sensor
BRET Study Part II
In vitro
In vivo
Prepare cell lysate in white opaque plate
Seed cells in black 96/384-well plate
Implant transduced cells in mice
Add substrate
Live cell
Inject Substrate
Immediate reading or incubate (depending upon substrate kinetics)
Add substrate
Image anesthetized animal
Quantitative estimation
Qualitative Quantitative estimation
Acceptor
Donor

donor and acceptor protein allows the fusion construct to fold properly and minimize the conformational constraints.

4. To ensure generation of fusion protein remove stop codon between the two cDNAs either using site-directed mutagenesis or delete using primer-based amplification.
5. To optimize suitable dipole orientation, while making BRET vectors, prepare all eight plasmid clones to find the best possible combination, i.e., pX-Rluc8.6 + pY-TurboRFP635 and pRluc8.6-X + pTurboRFP635-Y, pY-Rluc8.6 + pX-TurboRFP635 and pRluc8.6-Y + pTurboRFP635-X. An absence of BRET signal does not always mean that the two proteins are not interacting. It is possible that donor and acceptor dipoles are not aligned optimally to allow sufficient transfer of resonance energy. In this case clone donor–acceptor in both the orientation, i.e., N- and C-terminus, and select the appropriate orientation (*see* **Note 4**).

3.1.1 Fusion Construct Validation

1. Co-transfect cells with donor fusion construct and acceptor fusion construct that had the maximum BRET output signal. Simultaneously transfect cells with donor alone and keep a set of untransfected cells as well. Validate the BRET constructs either by preparing cell lysates or using live cells for imaging.
2. Fluorescence and luminescence study should also be done for transfected cells to judge the relative protein expression level. This is of particular significance while doing competition and saturation assay because the fluorescence and luminescence study will confirm that the changes in BRET signal obtained are not the result of the low expression of tagged proteins but the resultant of the assay [4].
3. Make 40–100 μl aliquots of sample/well in a 96-well plate, diluted in 1× PBS. Excite the fluorophore by using laser at specific wavelength, i.e., excitation wavelength of fluorophore in fluorometer/scanning spectrometer/flow cytometer, and collect the emission output signal at respective filter, i.e., emission wavelength of the fluorophore. For background correction of fluorescence, excite the untransfected cells also at the same wavelength followed by collecting the emission at respective filter. For example, TurboRFP635 has excitation maximum (Ex^{MAX}) at 588 nm and emission maximum (Em^{MAX}) at 635 nm.

Fig. 2 Diagrammatic illustration of basic steps involved in establishing BRET assay. BRET study 1 outlines the primary steps starting from vector selection to preliminary BRET methods required prior to in vitro or in vivo experiments. BRET study 2 highlights the different formats in which BRET assay can be performed using cell lysates, live cells, or animal model. For measuring donor and acceptor light output, a microplate reader can be used for experiments where only cell lysates or cultured live cells are involved, whereas CCD-coupled black box imaging device can be adopted for simultaneous quantitative and qualitative measurements at all levels. Bioluminescence imaging (BLI) microscopy or recently developed dual-view microimager with modified electron bombardment-CCD camera can also be used for live cell BRET measurement

4. Luminescence study can be done on same set of samples as used for fluorescence study. Add substrate, diluted in respective assay buffer, to the samples in 96-well plate and take readings in luminometer. Use only luciferase-expressing cell as positive and untransfected cell as negative control to compare luciferase activity. For background correction add substrate to untransfected cells as well and check luminescence signal, if any.
5. Check the functionality of target protein after fusion (as sometimes fusion tags might affect the biological function of the target protein) by performing respective functional assays of the protein, e.g., kinase assay, or localization studies using immunofluorescence/confocal microscopy [26].
6. Western blot should also be done to ensure that the fusion protein is fully translated by checking the size reference in the blot.
7. If one experiences cell death or change in cell growth pattern after transfection of BRET fusion plasmids, a cell viability assay using MTT [3-(4,5-dimethylthiazol-2-yl)-2,5-diphenyltetrazolium bromide] or trypan blue exclusion assay should be performed.

3.2 BRET for Protein–Protein Interaction Measurement

The technique of BRET has successfully been evolved to study protein–protein interaction in vivo and an increased BRET signal indicates the proximity achieved. However BRET does not provide a direct measure of physical interaction between the two proteins of interest. The same can be validated by various other assay formats like saturation assays, competition assays, and dilution assays [7].

3.2.1 BRET Saturation Assay

BRET saturation assay has successfully been used to demonstrate receptor oligomerization. It provides direct evidence for specific interaction as well as presence of physical contact between the two proteins in question. The saturation assay involves one protein tagged with donor molecule expressed in constant amount and subsequently increasing the amount of other protein tagged with the acceptor molecule. For specific interaction as the concentration of acceptor molecule increases the BRET signal will keep on increasing and attain saturation ($BRET_{max}$) level once all the donor molecules are occupied (hyperbola curve). Beyond $BRET_{max}$ any further increase in acceptor concentration will not enhance the BRET output signal. On contrary if the interaction is not specific or merely the result of random collision between the two proteins, then the BRET signal will continue to increase with increasing acceptor concentration in a quasi-linear fashion [7]. To perform BRET saturation assay, follow the steps below:

1. Seed cells in 6-well plate and culture for 24 h before transfection.
2. Transfect cells using appropriate transfection kit with constant amount of donor plasmid and increasing amount of acceptor

plasmid. In parallel maintain a set of donor-only transfected cells and untransfected cells for background correction.

3. Incubate at 37 °C and 5 % CO_2 for 24 h.
4. 48 h post-transfection, wash cells with 1× PBS (twice) and seed cells in 96-well plate (in duplicate or triplicate) at a density of ~100,000 cells/well for BRET assay.
5. Repeat **steps 3** and **4** of Subheading 3.1.1 for fluorescence and luminescence study.
6. Incubate at 37 °C for another 24 h to allow cells to adhere.
7. Dilute substrate in suitable medium/buffer and add to cells. Either take the BRET reading immediately or incubate cells depending upon the substrate kinetics using luminometer or IVIS.

3.2.2 Competition Assay

The specificity of protein–protein interaction can further be validated by BRET competition assay. In this assay the acceptor-tagged protein and donor-tagged protein are co-expressed along with an untagged protein or in the presence of an inhibitor that competes with one of the tagged proteins for interaction, at a single concentration (generally excess) or in a dose-dependent manner. If the two tagged proteins are specific interacting partners then with increasing concentration of the inhibitor or untagged competing protein the BRET signal will go down. In parallel also transfect an untagged noninteracting protein or a nonspecific inhibitor as a negative control that will not affect the BRET signal as it cannot interact with the tagged proteins [27]. The competition assay has been successfully used to study the oligomerization of GPCR receptors and dimerization of melatonin receptor [28].

1. Seed cells in 6-well plate and culture for 24 h before transfection.
2. Transfect cells using appropriate transfection kit with constant amount of donor plasmid and acceptor plasmid in the ratio of 1:1. In parallel maintain a set of cells transfected with noninteracting untagged protein plasmid at a single dose (excess) or increasing dose, untransfected cells, cells transfected with donor alone, and cells transfected with acceptor alone for background correction.
3. 48 h post-transfection, wash cells with 1× PBS (twice) and seed cells in 96-well plate (in duplicate or triplicate) at a density of ~100,000 cells/well.
4. Before taking BRET reading aliquot 40–100 μl sample and dilute in 1× PBS and 0.1 % BSA for fluorescence and luminescence analysis.
5. Repeat **steps 6** and **7** of Subheading 3.2.1.
6. If an antagonist is to be used in place of untagged interacting protein, after 48 h of transfection incubate ~10,000 cells in 1×

PBS and 0.1% BSA with a single excess dose of the radiolabeled antagonist or with the increasing concentration.

7. To check the nonspecific binding in negative control, treat cells with a radiolabeled nonspecific inhibitor that will not affect the interaction between tagged proteins.
8. Carry out the binding reaction at room temperature for 90 min and then stop the reaction by filtering through the Whatman/glass fiber filter.
9. The linear regression curve can be plotted between fluorescence and luminescence signal and total amount of tagged protein as determined using radiolabeled ligand binding in cells expressing each of the constructs individually.

3.3 In Vitro BRET Measurement from Cell Lysate

1. On day 1, seed cells in 6-well plate using the recommended complete culture medium and incubate at 37 °C, 5% CO_2. Typically there will be 50–80% confluency after 24 h of seeding. For example, 293 T or HT1080 cells are plated out at a density of $1.5–2 \times 10^5$ cells/well in 6-well plate. However, the amount of cells to be seeded will differ depending upon the growth rate, size of the cells, and the transfection kit protocol in use.
2. The following day, co-transfect cells with donor fusion construct and acceptor fusion construct, cloned in best orientation for optimal BRET signal, using suitable transfection reagent in accordance with the manufacturer's protocol. Depending on the form of analysis, a population of cells expressing only donor-labeled proteins at similar expression levels to those co-expressing donor- and acceptor-labeled proteins need to be generated. Also use suitable positive- and negative-control plasmids in parallel wells. Positive controls can be well-established interacting partners while the negative controls can be a biologically inactive mutant of one or both of the two protein partners. For background correction of fluorescence and luminescence, maintain a population of untransfected cells in parallel. Total amount of DNA used in transfection should be constant.
3. If interaction study of the two proteins of interest requires the presence of ligand or any other reagent, pre-treat the cells with the same at an appropriate time after transfection (after 24 h) and before BRET detection. Only vehicle-treated cells will serve as negative control.
4. The optimal expression time for transiently transfected proteins should be established before.
5. 24–48 h after transfection, aspirate medium from the cells and wash twice with 1× PBS. Trypsinize the cells and obtain cell pellet by centrifuging at 400*g*- force for 5 min, 4 °C.
6. Add 1× passive lysis buffer to the cell pellet at a volume approximately equal to thrice of the cell pellet.

7. Vortex to detach the pellet (if frozen, thaw on ice) after addition of lysis buffer and keep on ice for 20 min.
8. Collect supernatant by spinning the tubes at 2000*g*-force for 30 min, 4 °C.
9. Determine the protein concentration by mixing 5 μl of supernatant of each sample with 1 ml of 1× Bradford reagent. Mix well and take absorbance at 595 nm. Calculate the protein concentration using standard curve determined from known protein samples (e.g., BSA).
10. Prepare different dilutions of sample in 1× passive lysis buffer and aliquot 100 μl in 96-well plate. Add coelenterazine (1 μg/well concentration).
11. Take readings immediately on luminometer or microplate reader or IVIS Spectrum/IVIS 200.

3.4 BRET Measurement from Live Cell

Live cell protein–protein interactions are dependent on factors like subcellular localization, posttranslational modifications, and competitive interactions with other cellular partners. Currently used in vitro drug/ligand screening platforms are controlled and artificial, while for in vivo screening the drug/ligand should cross the cell plasma membrane, and reach their target protein in subcellular compartments with enough specificity to compete and interact exclusively with its target minimizing the potential interaction with thousands of other intracellular compounds. Thus the live cell protein–protein interaction studies are advantageous over classical in vitro biochemical analyses like co-immunoprecipitation, co-purification analysis, as in this case the host cell acts as a live cell test tube and allows noninvasive, quantitative, real-time readout of protein interactions in live cells even in single live cell format [21]. The live cell image can be performed in multiple formats as follows:

3.4.1 Live Cell BRET Measurement of Adherent Cells

1. Same as in Subheading 3.3, **steps 1–4**.
2. 24 h post-transfection, trypsinize the cells with trypsin–EDTA.
3. Resuspend cells in HEPES-buffered DMEM without phenol red and split the cells (in triplicate) at a density of 10,000–30,000 cells/100 μl/well in 96-well white cell culture plate, if measurement is being carried out by luminometer or into 96-well black cell culture plate, and if measurement is being carried out by IVIS.
4. Maintain cells at 37 °C, 5 % CO_2, in a humidified incubator for further 24 h before BRET assay, to allow cell attachment. To establish a suitable cell dilution initial titration is required.
5. Carry out the BRET assay in existing phenol red-free medium or replace medium with suitable assay buffer such as DPBS.

6. Remove medium from the cells and add substrate prepared in respective assay buffer. Following addition of coelenterazine h and 400a, BRET is determined immediately, while for EnduRen, incubate cells at 37 °C and 5% CO_2 in incubator for at least 1 h following addition of substrate and before proceeding for BRET (*see* **Note 5**).
7. Initialize the instrument and place the 96-well black culture plate inside the black box imaging chamber and close the door.
8. On IVIS software (Living Image software) set different parameters like imaging mode as luminescence, exposure time, binning (balances between sensitivity and resolution of the CCD camera, usually set at medium), emission filter (as per the acceptor emission wavelength), field of view (FOV, defines the size of the squares in the alignment grid), and subject height (0.5 cm for plate). Acquire spectral scan of the plate by selecting filters starting from 480 to 800 nm or above as per the acceptor emission wavelength. Acquisition time may vary from seconds to minutes depending upon the reporter and the substrate kinetics. If an interaction has to be monitored at different time intervals, the IVIS instrument has the facility of setting up delay time for spectral scan ranging from seconds to minutes. To acquire image make sure that the "Photograph" and "Overlay" (to obtain a co-registered image) buttons are in on mode. Then go to image setup and select respective emission filters and acquire the sequences by clicking "acquire sequence" button or set time points for spectral scan in delayed kinetic assays.

   If fluorescent needs to be measured, set imaging mode in fluorescent, and the excitation filter is automatically selected based on the wavelength of the emission filter selected from the IVIS System control panel. However, the automatic selection in the IVIS System control panel can be overridden.
9. At first the camera acquires the photograph of the plate followed by the luminescence for the set period of time. As soon as the acquisition is over, a superimposed image of the photograph and pseudocolor luminescence image will appear on the screen.
10. After image acquisition, save the image data to desired location. For data analysis, draw ROI on the target sites; the measured photon values expressed as photons/s/cm^2/sr (steradian: a measure of solid angle) will be displayed in a new window that can be exported to Microsoft Excel for further use and statistical analysis. However, the total photons (photon/s) from a specific ROI can also be used for the analysis.
11. Calculate BRET ratios as described in Subheading 3.6.

3.4.2 Live Cell BRET Measurement of Cell Suspensions

1. Same as in Subheading 3.3, **steps 1–4**.
2. Typically 48 h after transfection, detach the cells using trypsin–EDTA.

3. Resuspend cells in suitable BRET assay buffer at required dilution and plate in 96-well white cell culture plate or 96-well black cell culture plate depending on the instrument used.
4. Same as in Subheading 3.4.1, **steps 7–11**.

3.4.3 BRET Kinetic Measurement

General substrates for Rluc-based BRET assays are coelenterazine h or DeepBlueC™. However these substrates have major drawbacks like less stability in aqueous solution at physiological temperature (37 °C) and enhanced autofluorescence in the presence of serum that limits its use for assays with prolonged kinetics. A new or modified version of coelenterazine, EnduRen™, has been produced that can be activated only and after being acted upon by cellular esterase in live cells to produce free coelenterazine h. Once coelenterazine h is produced it can interact with the respective donor molecule to produce the BRET output signal. This version of BRET measurement of dynamic events was named as eBRET or extended BRET that utilizes EnduRen™ as luciferase substrate (*see* **Note 6**). The method provides potential advantage of monitoring protein–protein interaction in live cells under physiological conditions for prolonged hours without significant depletion of the output signal [18].

The IVIS Spectrum/IVIS 200 optical imaging systems provides option of performing BRET assay using sequential mode wherein image sequences can be collected at delayed time frame defined by user. Using this feature, it is possible to capture the kinetics of a protein–protein interaction assay in a real-time manner.

3.4.4 Live Cell Imaging Using BLI Microscope

1. Seed the cells in 6-well plate.
2. Transient transfection or stable cell generation with respective plasmids.
3. Addition of respective ligand and substrate. Allow substrate incubation at 37 °C, 5% CO_2, up to 2 h, if the substrate is EnduRen™.
4. Image cells with an Olympus LV200 luminescent microscope (Olympus America, Inc., Melville, NY, USA) with respective donor filter and acceptor filters.
5. Adjust the acquisition time to resolve the cells clearly.
6. For measuring mean integrated pixel densities on regions of interest software like ImageJ (NIH, Bethesda, MD, USA) can be used.
7. Do the ratiometric calculation of acceptor/donor signal (refer to Subheading 3.6).

3.4.5 Multiplexed BRET Assay Using Spectral Imaging

This can be performed either from live cell or cell lysate. Protocol for assay setup is same as described in Subheading 3.4.2, except that the scan has to be performed using IVIS Spectrum or IVIS200 loaded with 20 nm spectral band-pass filter ranging from 460 to 800 nm. Using this equipment, it is possible to set a sequential scan using emission filter sets ranging from 480 nm to 700 nm (that is where most of the current BRET donor and acceptor elements emit). Total spectral scan time may vary according to reporter intensity, but generally require 2–5 min.

1. Ideally for spectral scan procedure one should use substrates that are soluble, stable at 37 °C, and protected in live cell to yield stable signal intensity for certain time or during the course of the assay. These properties of substrate allow for BRET spectral studies of cellular functions at extended time scale.
2. To control signal intensity variation, the scan sequence should collect image using total light at the beginning and at the end of the spectral images. It should be assured that the total light output remain unaltered during the spectral scan. Any decrease or increase in the total light output will affect or produce error in the calculation of BRET ratio for protein interaction assay.

3.5 In Vivo BRET Measurement from Animal Model

3.5.1 Cell Implantation

For short-term protein–protein interaction studies either use transiently transfected cells or cells stably expressing the fusion construct: (1) donor alone, (2) positively interacting fusion constructs, and (3) negatively interacting fusion constructs. A maximum of four sites can be used in single-animal model for implanting cells. On average cells between one and five million can be implanted in a single animal.

To study the ligand-dependent protein–protein interaction in living animals, the mice is first injected with suitable concentrations of the ligand (reagent) through tail vein while the control mice receives the same volume of vehicle.

3.5.2 Animal Anaesthesia

Before conducting any animal experiment a project license must be obtained from both the local and national animal ethics committee that ensures that all ethical concerns are addressed prior to conducting animal experiments. All experimentation protocols should be ethical and humane and only a well-trained person should be allowed to handle and experiment on live animals. It is also important that the individual must follow all the local and national guidelines set up for ethical use and care of animals during performing an experiment.

Intraperitoneal (i.p.) injection (40 μl/25 g body weight) of ketamine and xylazine solution at a ratio of 4:1 can be used for anesthetizing mouse. This method can be used for imaging experiments lasting up to 30 min. For experiments where repeated mouse imaging within a day is required, try to use isoflurane gas anesthesia (*see* **Note 7**).

3.5.3 Instrument Setup

In the meantime initialize the IVIS instrument. Adjust different parameters same as in Subheading 3.4.1, **step 8**. For animal imaging set subject height at 1.5 cm. FOV is adjusted at 25 cm^2 for five mice and 10 cm^2 for one mice, and it varies depending upon the number of mice to be scanned at a time.

3.5.4 Substrate Delivery

1. Preferred location: Lateral veins on animal tail are the preferred location for i.v. tail vein injection.
2. Needle used for mice: sterile small 28–30 G, usually used with 1 cc insulin syringe.
3. A dose of 1 mg/kg body weight of coelenterazine is recommended. For example, for a mouse weighing 20 g, inject 200 μl of 1 μg/10 μl to deliver 20 μg of coelenterazine. Higher coelenterazine concentration up to five times (100 μg) can be required for some applications (*see* **Note 8**).
4. Inject the substrate via intravenous (i.v.) tail vein route and image immediately by placing the subject inside the imaging platform (*see* **Note 9**).
5. Injection method: Under deep anaesthesia place the animal on side. Grasp the tail at the distal end. Place the index and middle fingers of the non-dominant hand around the tail above the site of needle insertion (these fingers act as a tourniquet) and the lower part of the tail is held between the thumb and ring fingers below the injection site. Slight opposing pressure is applied with both sets of fingers to straighten and stabilize the tail. Needle should be level-side up and slightly angled when entering the veins. It should be advanced parallel to the vein approximately ½ (~5 mm) of the tail length; protrude the needle into the vein being very careful not to perforate the vein. Draw back on the syringe slightly and look for traces of blood flow into the needle hub indicating that the needle is successfully inserted into the vein. Release pressure before administering the substrate steadily over few seconds into the vein. There should be minimal resistance during injection. Remove the needle and apply gentle compression until bleeding stops and perform scan by placing the animal within the field of view of the imager. Return animals to their cage and observe for 5–10 min to make sure that bleeding has stopped (*see* **Notes 10–12**).

3.5.5 Image Acquisition in Dark Chamber Using Cooled CCD Camera in IVIS

1. Place the anesthetized animal inside the temperature-controlled lighttight black box imaging chamber, either prone or supine, depending on the site of cell implantation or tumor growth and close the door. For example, if the implanted cells are on back, place the animal exposing dorsal side towards the camera

so that the path length of fluorescent light through different tissues or organs is minimized.

2. Acquire, save, and analyze data as described in Subheading 3.4.1, **steps 9–11**.

3.6 BRET Data Analyses

3.6.1 BRET Ratio Calculation and Interpretation

1. BRET ratio is calculated as the "emission through the acceptor wavelength filter" divided by the "emission through the donor wavelength filter." For example, in BRET[8] with TurboRFP as acceptor, emission through 630 nm filter over 540 nm filter is to be measured.
2. Measurement of BRET ratio in in vitro and in vivo studies is carried out using the following generalized equation [3]:

$$BRET\ ratio = \frac{BL\ emission\left(Acceptorl\right) - cf \times BL\ emission\left(Donor\ \lambda\right)}{BL\ emission\left(Donor\ \lambda\right)}$$

$$Cf = \frac{\mathrm{BL}_{\mathrm{emission}}\left(\mathrm{Acceptor}\ \lambda\right)_{\mathrm{donor\ only}}}{\mathrm{BL}_{\mathrm{emission}}\left(\mathrm{Donor}\ \lambda\right)_{\mathrm{donor\ only}}}$$

where BL is the average radiance and *Cf* is the correction factor.

3. To study ligand- (reagent-) induced protein–protein interaction, BRET data are collected prior to addition of ligand (reagent). Then ligand (reagent) is added preferably through an injector (available with luminometer, BMG Labtech), if post-addition early time points (<1 min) are needed. Repeated measurements are taken over a period of time to determine the effect. To provide the control for background signal, vehicle-treated samples are measured in parallel.
4. The BRET ratio measurement for ligand-induced (reagent) interaction involves BRET ratio from both ligand (reagent)-treated and vehicle-treated samples (samples can be cells or cell lysates from cells co-expressing donor and acceptor fusion proteins). BRET ratios of both ligand (reagent)-treated and vehicle-treated samples need to be subtracted first from BRET ratio of untransfected cells. Then subtract the BRET ratio of vehicle-treated samples from the ligand (reagent)-treated samples which gives us the "ligand (reagent)-induced BRET ratio" (*see* **Note 13**).
5. If the ligand treatment results in a negative BRET ratio, it may imply that there are no or weaker interactions and/or more transient interactions than those observed prior to ligand addition. This may be the case of conformational change of the interacting proteins resulting in the greater distance between them or less favorable relative orientation.

6. The BRET signal can be plotted against time to obtain a time kinetics profile from which apparent association (or dissociation) rate constants can be find out.
7. Ratiometric analysis of PPIs by a BRET in tissue background is hindered considerably by higher tissue attenuation for shorter wavelength light as compared to longer wavelength light (especially >600 nm), which is mainly associated with absorption by hemoglobin and myoglobin [29]. To ensure that the BRET ratio remains constant between cultured cells and mice, the imaging results may be analyzed using the double-ratio (DR) method [30], which partially corrects for signal attenuation (*see* **Note 14**).

   The equation for calculating double ratio is as follows:

$$DR = \frac{\dfrac{BL_{\text{emission}}\left(\text{Acceptor }\lambda\right)_{\text{BRET}}}{BL_{\text{emission}}\left(\text{Donor }\lambda\right)_{\text{BRET}}} \quad \dfrac{\mu t\left(\text{Acceptor }\lambda\right)}{\mu t\left(\text{Donor }\lambda\right)}}{\dfrac{BL_{\text{emission}}\left(\text{Acceptor }\lambda\right)_{\text{donor only}}}{BL_{\text{emission}}\left(\text{Donor }\lambda\right)_{\text{donor only}}} \quad \dfrac{\mu t\left(\text{Acceptor }\lambda\right)}{\mu t\left(\text{Donor }\lambda\right)}}$$

   which is independent of μt (total attenuation coefficient).

4 Notes

1. Coelenterazine is prone of self-oxygenation in the presence of light and at higher temperature. Therefore, always store coelenterazine in dark tubes stored at −80 °C freezer. For running use, store at −20 °C.
2. Dilute the substrates in appropriate assay buffers immediately before adding to samples and protect from light. For EnduRen to avoid precipitation, preincubate the assay buffer at 37 °C.
3. Black mice show 10× reduction in bioluminescence and 20× reduction in fluorescence signal. Hence it is recommended to restrict experiments to the use of nude or white fur mice.
4. Apart from dual-vector construction for donor and acceptor proteins, a single-vector format can also be used in which both acceptor and donor proteins are cloned along with the interacting protein partners in a single-vector backbone (e.g., pRluc8.6-X-Y-TurboRFP635 or pRluc8.6-Y-X-TurboRFP635). Previously De A et al. have demonstrated construction of such single-vector (*pCMV-mOrange-FRB-FKBP12-RLuc8*) plasmid [21].
5. Washing is generally not recommended when conducting experiments in live cell format as cells may get washed off from well causing difference in light output. Many cell types are loosely

attached on the plate; hence extreme care should be taken during medium aspiration, if any.

6. Activation of EnduRen™ requires activity of cellular esterases; hence EnduRen™ can only be used for live cell BRET imaging and not for in vitro BRET studies.
7. Anesthetic drugs are highly regulated products, available under licensed prescription only. Stage temperature should be maintained at 37 °C, to avoid drop in the animal body temperature during and after imaging experiments, until the animal returns to conscious state.
8. Coelenterazine is prone to self-oxidation at room temperature and light; thus working solutions must be kept until use in ice and light-protected condition.
9. As coelenterazine has flash time kinetics and signal decays very quickly, a coelenterazine kinetic study should be performed for each animal immediately after substrate administration. Generally tolerated volume of i.v. injection is up to 200 μl of aqueous solution.
10. Great care needs to be taken during the tail vein injection. A successful tail vein flushes out the red color of the vein while pushing the injection and a blood droplet oozes out after the needle is withdrawn. If the vein is missed during injection, the substrate will then go into the surrounding subcutaneous and dermal tissues, resulting in a blanching and bulging at the injection site. If this occurs, carefully withdraw the needle and reattempt injection at a more proximal (towards the mouse's body) location.
11. Body temperature of the animal must be maintained at 37 °C during i.v. injection, preferably by means of a heated stage or heating lamp.
12. Care should be taken not to spill coelenterazine during injection as any spillage of substrate at the site of injection can give rise to strong background signal due to autofluorescence of coelenterazine.
13. For ligand-induced interactions donor-only controls are not needed as vehicle-treated samples will represent the background. Moreover, since BRET-based assays are ratiometric, any variability due to assay volume or cell number variation or time point of measurement is nullified.
14. DR is a dimensionless parameter independent of the total attenuation coefficient assuming that the attenuation coefficient is constant for all mice and identical over the entire thorax area. The DR method provides a depth and number of reporter cell-independent measure of the BRET signal; however, both donor and acceptor signals used to calculate the DRs decrease with tissue depth.

Acknowledgements

Research funding (BT/PR3651/MED/32/210/2011) from Department of Bioengineering, New Delhi, India, to A.D. is acknowledged.

References

1. Ngounou Wetie AG, Sokolowska I, Woods AG et al (2013) Investigation of stable and transient protein–protein interactions: past, present, and future. Proteomics 13:538–557
2. Stryer L, Haugland RP (1967) Energy transfer: a spectroscopic ruler. Proc Natl Acad Sci U S A 58:719–726
3. De A, Jasani A, Arora R et al (2013) Evolution of BRET biosensors from live cell to tissue-scale imaging. Front Endocrinol (Lausanne) 4:131
4. Pfleger KD, Seeber RM, Eidne KA (2006) Bioluminescence resonance energy transfer (BRET) for the real-time detection of protein–protein interactions. Nat Protoc 1: 337–345
5. Drinovec L, Kubale V, Nohr LJ et al (2012) Mathematical models for quantitative assessment of bioluminescence resonance energy transfer: application to seven transmembrane receptors oligomerization. Front Endocrinol (Lausanne) 3:104
6. Subramanian C, Xu Y, Johnson CH et al (2004) In vivo detection of protein–protein interaction in plant cells using BRET. Methods Mol Biol 284:271–286
7. Ayoub MA, Pfleger KD (2010) Recent advances in bioluminescence resonance energy transfer technologies to study GPCR heteromerization. Curr Opin Pharmacol 10:44–52
8. Perroy J, Pontier S, Charest PG et al (2004) Real-time monitoring of ubiquitination in living cells by BRET. Nat Methods 1:203–208
9. Kim YP, Jin Z, Kim E et al (2009) Analysis of in vitro SUMOylation using bioluminescence resonance energy transfer (BRET). Biochem Biophys Res Commun 382:530–534
10. Schroder M, Kroeger KM, Volk HD et al (2004) Preassociation of nonactivated STAT3 molecules demonstrated in living cells using bioluminescence resonance energy transfer: a new model of STAT activation? J Leukoc Biol 75:792–797
11. Deplus R, Delatte B, Schwinn MK et al (2013) TET2 and TET3 regulate GlcNAcylation and H3K4 methylation through OGT and SET1/COMPASS. EMBO J 32:645–655
12. Dionne P, Mireille C, Labonte A, Carter-Allen K, Houle B, Joly E, Taylor SC, Menard L (2002) BRET2: Efficient energy transfer from Renilla Luciferase to GFP2 to measure protein–protein interactions and intracellular signaling events in live cells. In: van Dyke K, van Dyke C, Woodfork K (eds) Luminescence biotechnology: instruments and applications. CRC, Boca Raton, FL, pp 539–555
13. Loening AM, Fenn TD, Wu AM et al (2006) Consensus guided mutagenesis of Renilla luciferase yields enhanced stability and light output. Protein Eng Des Sel 19:391–400
14. Hall MP, Unch J, Binkowski BF et al (2012) Engineered luciferase reporter from a deep sea shrimp utilizing a novel imidazopyrazinone substrate. ACS Chem Biol 7:1848–1857
15. Inouye S, Shimomura O (1997) The use of Renilla luciferase, Oplophorus luciferase, and apoaequorin as bioluminescent reporter protein in the presence of coelenterazine analogues as substrate. Biochem Biophys Res Commun 233:349–353
16. Otsuji T, Okuda-Ashitaka E, Kojima S et al (2004) Monitoring for dynamic biological processing by intramolecular bioluminescence resonance energy transfer system using secreted luciferase. Anal Biochem 329: 230–237
17. Remy I, Michnick SW (2006) A highly sensitive protein–protein interaction assay based on Gaussia luciferase. Nat Methods 3:977–979
18. Pfleger KD, Dromey JR, Dalrymple MB et al (2006) Extended bioluminescence resonance energy transfer (eBRET) for monitoring prolonged protein–protein interactions in live cells. Cell Signal 18:1664–1670
19. Xie Q, Soutto M, Xu X et al (2011) Bioluminescence resonance energy transfer (BRET) imaging in plant seedlings and mammalian cells. Methods Mol Biol 680:3–28
20. De A (2011) The new era of bioluminescence resonance energy transfer technology. Curr Pharm Biotechnol 12:558–568
21. De A, Ray P, Loening AM et al (2009) BRET3: a red-shifted bioluminescence resonance

energy transfer (BRET)-based integrated platform for imaging protein–protein interactions from single live cells and living animals. FASEB J 23:2702–2709

22. Xu Y, Piston DW, Johnson CH (1999) A bioluminescence resonance energy transfer (BRET) system: application to interacting circadian clock proteins. Proc Natl Acad Sci U S A 96:151–156
23. De A, Loening AM, Gambhir SS (2007) An improved bioluminescence resonance energy transfer strategy for imaging intracellular events in single cells and living subjects. Cancer Res 67:7175–7183
24. De A, Arora R, Jasani A (2014) Engineering aspects of bioluminescence resonance energy transfer systems. In: Cai W (ed) Engineering in translational medicine. Springer, London, pp 257–300
25. Xu X, Soutto M, Xie Q et al (2007) Imaging protein interactions with bioluminescence resonance energy transfer (BRET) in plant and mammalian cells and tissues. Proc Natl Acad Sci U S A 104:10264–10269
26. Kocan M, Pfleger KD (2011) Study of GPCR-protein interactions by BRET. Methods Mol Biol 746:357–371
27. Couturier C, Deprez B (2012) Setting up a bioluminescence resonance energy transfer high throughput screening assay to search for protein–protein interaction inhibitors in mammalian cells. Front Endocrinol (Lausanne) 3:100
28. Ayoub MA, Couturier C, Lucas-Meunier E et al (2002) Monitoring of ligand-independent dimerization and ligand-induced conformational changes of melatonin receptors in living cells by bioluminescence resonance energy transfer. J Biol Chem 277:21522–21528
29. Zhao H, Doyle TC, Coquoz O et al (2005) Emission spectra of bioluminescent reporters and interaction with mammalian tissue determine the sensitivity of detection in vivo. J Biomed Opt 10:41210
30. Dragulescu-Andrasi A, Chan CT, De A et al (2011) Bioluminescence resonance energy transfer (BRET) imaging of protein–protein interactions within deep tissues of living subjects. Proc Natl Acad Sci U S A 108:12060–12065

Chapter 6

Studying Nuclear Receptor Complexes in the Cellular Environment

Fred Schaufele

Abstract

The ligand-regulated structure and biochemistry of nuclear receptor complexes are commonly determined by in vitro studies of isolated receptors, cofactors, and their fragments. However, in the living cell, the complexes that form are governed not just by the relative affinities of isolated cofactors for the receptor but also by the cell-specific sequestration or concentration of subsets of competing or cooperating cofactors, receptors, and other effectors into distinct subcellular domains and/or their temporary diversion into other cellular activities. Most methods developed to understand nuclear receptor function in the cellular environment involve the direct tagging of the nuclear receptor or its cofactors with fluorescent proteins (FPs) and the tracking of those FP-tagged factors by fluorescence microscopy. One of those approaches, Förster resonance energy transfer (FRET) microscopy, quantifies the transfer of energy from a higher energy "donor" FP to a lower energy "acceptor" FP attached to a single protein or to interacting proteins. The amount of FRET is influenced by the ligand-induced changes in the proximities and orientations of the FPs within the tagged nuclear receptor complexes, which is an indicator of the structure of the complexes, and by the kinetics of the interaction between FP-tagged factors. Here, we provide a guide for parsing information about the structure and biochemistry of nuclear receptor complexes from FRET measurements in living cells.

Key words Förster resonance energy transfer, Cellular biochemistry, Protein structure, Fluorescence microscopy, Androgen receptor

1 Introduction

Appending the cDNA for any protein with the open reading frame of a fluorescent protein (FP) cDNA allows the FP-labeled protein to be expressed in cells and tracked by fluorescence microscopy [1–4]. Beyond simply measuring the cellular locations of the FP-tagged factors, a number of techniques are available that provide more detailed information about the FP-tagged proteins and their complexes in the cellular environment [5–8]. Those methods include (1) fluorescence correlation spectroscopy (FCS), in which the duration and amount of fluorescent emissions from an FP-tagged complex within a small cellular volume provide information about the size

Iain J. McEwan (ed.), *The Nuclear Receptor Superfamily: Methods and Protocols*, Methods in Molecular Biology, vol. 1443,
DOI 10.1007/978-1-4939-3724-0_6,

and concentrations of the complex [9–11]; (2) fluorescence recovery after photobleaching (FRAP) and related techniques (iFRAP, FLIP, or photoactivation) that distinguish rapidly diffusing and stable subpopulations of FP-tagged complexes at subcellular locales by the movements of FP-tagged complexes into previously photobleached cellular regions (FRAP), from cellular regions near to continuously photobleached sites (iFRAP and FLIP) or from sites of fluorescence photoactivation [9, 12, 13]; (3) bimolecular fluorescence complementation (BiFC) in which stable (~1 h) interactions between two factors tagged with complementary, non-fluorescent "halves" of an FP can enable those halves to form a fluorescent pseudo-FP [14]; and (4) Förster resonance energy transfer (FRET), or the related bioluminescence resonance energy transfer (BRET), in which transient or stable interactions are detected between two factors labeled with different FPs (or with one FP and one luminescent protein), one of which has an emission energy that overlaps with the excitation energy for the other [15–19]. Many of the above techniques can be combined into very powerful approaches for discerning the properties of FP-tagged molecules in living cells [20, 21]. Some reviews and representative publications [22–40] provide examples of the extensive application of these techniques to studies of nuclear receptors in the cellular milieu.

General concepts needed to understand FRET measurement will be provided in this Introduction section (Fig. 1), which will include some examples of the measurements to be detailed in Subheadings 3 and 4. A prior chapter in this series provided more details of FRET measurement itself [41], which is only summarized here. This chapter focuses on techniques that examine subcellular and cell-to-cell differences in FRET for cataloging natural variations in biologic response (Fig. 2) and for extracting information about the biochemistry and structure of interacting factors from logical patterns within that measurement variability (Fig. 3).

Figure 1 depicts the fundamental concept of FRET. A fluorophore (Fig. 1a, e.g., CFP = cyan FP) becomes activated through absorption of a photon (lightning bolt) of a specific energy level (wavelength) required to activate CFP. That activation results in the formation of a transient energy field (Fig. 1b). In the absence of any other influences, the activated FP returns to its ground state, in part, through the release of a photon (Fig. 1c). The emitted energy is lower (longer wavelength) than the energy of the activating photon since some energy in the activated state is dissipated through other means. If another FP happens to come within the transient energy field of the activated "donor" FP (Fig. 1d), that energy might be transferred to the second "acceptor" FP, but only if the energy within the donor field is of a level that can excite the acceptor FP.

Figure 1e depicts the energy overlap requirement as a shaded area required for acceptor FP excitation and the wavelengths of light that would be emitted by the donor FP. However, the excited donor

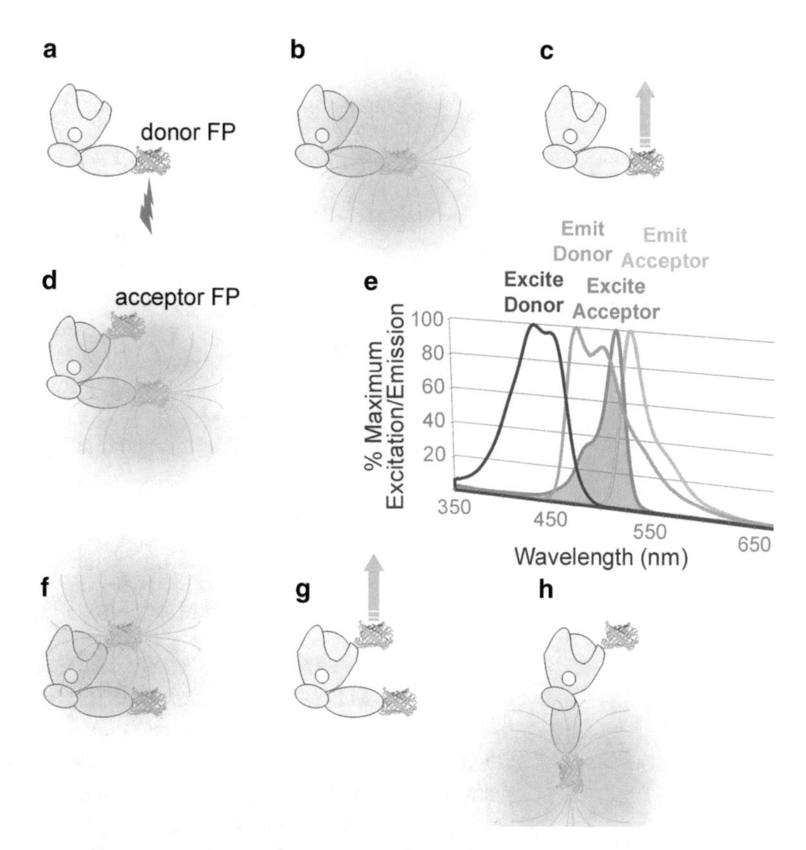

Fig. 1 Förster resonance energy transfer. A "donor fluorophore" tagged to a nuclear receptor and excited by capture of a photon of energy (lightning bolt) (**a**) creates a transient energy field (**b**) that may be dissipated by the emission of a photon of light characteristic of the donor (**c**). Alternatively, if an "acceptor fluorophore" is within the donor energy field (**d**) and capable of being excited by the energy within the donor field (**e**), then donor energy can transfer to the acceptor fluorophore (**f**) whereupon it emits a photon of light characteristic of the acceptor (**g**). If the acceptor is not within the donor energy field (**h**), no energy transfer occurs

FP never actually emits a photon during energy transfer. Instead, the energy within the activated donor FP is transferred "non-radiatively" to the nearby acceptor FP (*see* **Note 1**: distance influence). The activated acceptor FP (Fig. 1f) subsequently emits a photon of light characteristic of the acceptor FP (Fig. 1g). Thus, FRET is accompanied by a loss in donor FP emission intensity and an increase in acceptor FP emission intensity upon excitation of the donor FP, which is what is measured by the "sensitized emission" measurements detailed in this chapter. Other methods for measuring FRET not discussed here include (a) fluorescence lifetime imaging microscopy which detects the FRET-dependent decrease in the average time it takes for the donor FP to emit a donor photon [42, 43] and (b) fluorescence anisotropy imaging in which the anisotropy of the donor FP is reduced upon energy transfer [44, 45]. Each of the

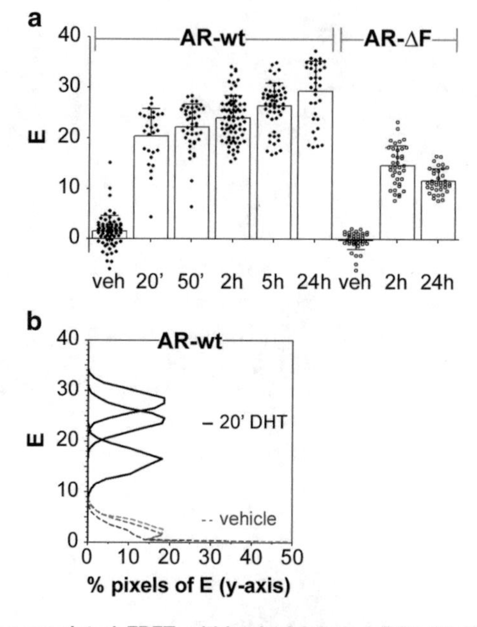

Fig. 2 Androgen-regulated FRET within dual-labeled CFP-AR-YFP reporter. (**a**) Percentage of CFP energy transferred to YFP at indicated times following the addition of 10^{-7}M DHT to the cells. Each *dot* represents E measured in an individual cell nucleus; the means ± standard deviations for all nuclei at each time point are shown in *bar graphs*. *wt* wild-type AR, *ΔF* AR deleted of five amino acids (FQNLF) in amino terminal domain. (**b**) Pixel-by-pixel frequency distribution of FRET throughout the nuclei of representative cells

FRET measurement methods has its advantages and, if conducted accurately, provides the same measurement on the same FP-labeled probes [46, 47]. Lifetime and anisotropy microscopes are not present in most laboratories, which is why FRET is more commonly conducted by sensitized emission measurements with intensity-based fluorescence microscopes. Regardless of how FRET is measured, the principles about extracting information from the variations in FRET measurement discussed in this chapter still hold.

Because FRET requires the presence of the acceptor FP in the extremely small energy field of the donor FP (typically <80 Å from the center of the donor; compared to the ~30–50 Å diameter of a typical single domain within a protein), donor and acceptor FPs are seldom in solution at sufficient concentrations for energy transfer to occur. FRET therefore occurs when the researcher positions the FPs close to each other either by placing two FPs on the same protein or by attaching them to interacting proteins that bring the FPs together when the complex forms (*see* **Note 2**: membrane effects). In either instance, the FPs must be positioned close enough so that the energy fields overlap [48, 49]. If the acceptor FP is not within the energy field of the donor, no energy transfer can occur even if they are attached to a single protein (Fig. 1h).

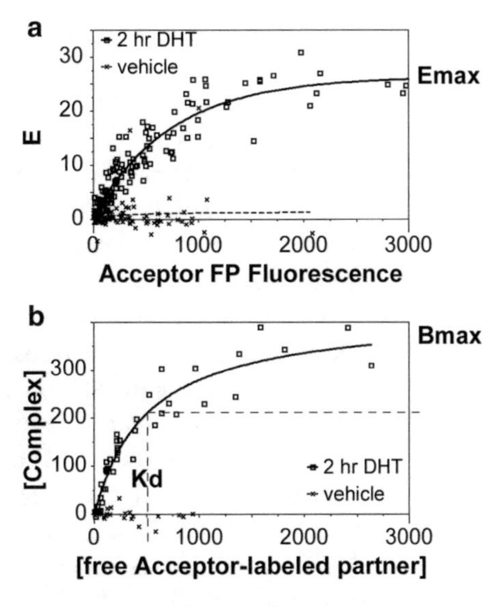

Fig. 3 Cellular biochemistry. (**a**) Relationship between the amount of energy transfer, E, to the amount of acceptor FP-labeled factor in the cell. Emax, maximal level of energy transfer when essentially all available donor FP-labeled factor is bound with acceptor FP-labeled factor. The determination of the Emax allows estimation of the proportion of donor-labeled factor bound at any data point as E/Emax which permits (**b**) the calculation of the amounts of fluorescence units of complex formed (*y*-axis) in relationship to the levels of unbound acceptor-FP-labeled factor, as described by the law of mass action. Note that the fluorescence units of the donor and acceptor FPs are described in molar equivalent units through the calibration procedures described in this chapter. They can easily be converted into molar concentrations when an instrument has been calibrated to convert fluorescence intensities as factor concentrations [55]

The sensitized FRET measurement relies, in part, on quantifying acceptor FP emissions following donor FP excitation. However, that is overly simplistic as FRET emissions in that measurement condition are tiny compared to strong fluorescence of the donor and acceptor FPs themselves under those same conditions. FRET measurement thus requires calibration for the removal of these "bleed-through" emissions of the acceptor and donor FPs from the measurement. Other calibrations also are necessary to express the amount of FRET as the percentage of donor FP emissions transferred to acceptor FP (E) [50, 51]. For interpreting FRET data, the amount of E quantified will be affected by two major influences [48, 49, 52]: (a) the position of the acceptor fluorophore within the energy field (example: deep, as in Fig. 1d would lead to a high E, which falls to the sixth power with the distance of the acceptor away from the donor until, when beyond the periphery of the energy field as in Fig. 1h, no energy transfer is possible) and (b) the relative orientations of the donor and acceptor fluorophores (energy transfer is most efficient when the energy fields are optimally aligned) (*see* **Note 3**: orientation influences). Thus, the amount of energy transfer is governed by the distance between, and the relative orientations of, the FPs within the protein (Fig. 1), which provides as a surrogate indicator of protein conformation.

In the example shown (Fig. 2a), an androgen receptor (AR-wt, wild-type) was labeled with two FPs (as in Fig. 1d–h) and expressed transiently in cells. Treatment with an agonist ligand (10^{-7} M dihydrotestosterone, DHT) resulted in a time-dependent change in AR conformation, reflected by changes in E. Each data point represents the amount of energy transfer averaged throughout a single-cell nucleus. The mean ± standard deviation for all cell nuclei at each time point is shown in the surrounding bar graphs. There is a rapid increase in E following DHT addition (compare 20′ with vehicle control), followed by a prolonged gradual increase after that. Mutation of the FQNLF motif in the AR amino terminus (Fig. 2a, AR–ΔF), which can interact with activation function-2 in the agonist-bound AR ligand-binding domain [53], reduces the amount of FRET detected and eliminates the further gradual increases in FRET at later stages of chronic ligand stimulation [28, 54].

Figure 2a depicts also the cell-to-cell heterogeneity in FRET measurement which may reflect (a) cell-to-cell differences in AR conformation and/or (b) errors in E determination. Measurement error is relatively low, as best depicted by vehicle-treated cells expressing the ΔF mutant ($E = 0.0\% \pm 1.8\%$). Even prior to the addition of DHT, ~5% of cells expressing the wild-type AR already display a somewhat "folded" AR. At the earliest stages (vehicle, 20′ and 50′) after ligand addition, statistical analysis shows the cell-to-cell variations in FRET level to be non-normal whereas at all time points of 2 h and above, the distribution in FRET level becomes normal (*see* **Note 4**: tracking cells at different time points). What about site-to-site variations in FRET level within a cell nucleus? The smallest unit upon which FRET can be calculated by digital fluorescence microscopy is a pixel. Figure 2b shows the proportion of pixels (*x*-axis) with each FRET level (*y*-axis) for representative cells treated with vehicle or treated with ligand for 20 min (all cells in the study were examined with similar results, not shown). In the absence of ligand, many cells have a small proportion of pixels ("intracellular sites") with a low level of FRET. However, even after 20 min of agonist addition, most cells show a characteristic FRET level that is normal around a mean. Such methods suggest that there is a subpopulation of sites in many unliganded cells where the AR is poised for action and that, upon the addition of ligand, the AR attains a conformation that is generally similar throughout each cell nucleus (*see* **Notes 5** and **6**: interpreting FRET at individual pixels).

Variations in FRET levels also represent the key to defining protein biochemistry measured by FRET. When attached to two different factors, the likelihood that the donor FP will transfer energy is a function of the likelihood of interactions that bring the

FPs into close enough proximity. Thus, when FRET depends upon interaction between FP-tagged factors, interaction kinetics will cause a predictable variance in measurement that will be detailed in this chapter. Figure 3a illustrates this concept in a study of the interaction between two ARs, one tagged with CFP (donor) and the other tagged with YFP (acceptor) (*see* **Note 7**: oligomerization). AR generally does not oligomerize (i.e., no FRET) when the cells are not treated with DHT (x in Fig. 3a), although 2–3 % of untreated cells showed evidence of sporadic AR:AR interaction again suggesting that cell-to-cell heterogeneity may be a component of biologic response. Following 2 h of incubation with 10^{-7}M DHT (open squares), the amount of energy transfer follows a pattern that increases with the amount of acceptor to which the donor can transfer energy to. This data can be fit to a curve (Fig. 3a) that, as summarized in [55] and in this chapter, permits extrapolation of the data into units reflecting the concentrations of the complex (Fig. 3b, *y*-axis) and the "unbound" acceptor FP-labeled factor (Fig. 3b, *x*-axis). The curve in Fig. 3b depicts an interaction between two factors described by the law of mass action, and permits the extrapolation from FRET data of standard biochemical parameters, including Bmax and Kd plus a novel "availability" factor unique to interactions measured in the complex cell environment.

2 Materials

2.1 Cell Preparation and Growth

1. Microscope slides and No. 1 borosilicate cover glasses. *See* **Note 8** for chamber and plate alternatives to the simple slide procedure.
2. Appropriate cell line in which to study AR function.
3. Expression vectors for FP-tagged proteins (*see* **Note 9**) that retain their function when fused to an FP (*see* **Note 10**).
4. Tissue culture media (*see* **Note 11**) appropriate for the cell line, supplemented with charcoal-stripped serum depleted of androgens.
5. Transfection reagent appropriate for cell line (*see* **Note 12**).
6. AR agonist (e.g., DHT, testosterone) or antagonist (e.g., Casodex, MDV3100, ARN-509) ligands.

2.2 Image Collection, Background Subtraction, Segmentation, and Quantification

1. Excitation and emission filters and dichroic mirrors to enable the collection of acceptor, donor, and FRET channels (*see* **Note 13**).
2. Software to control microscope and enable rapid image capture (*see* **Note 14**).
3. Quantitatively linear camera (*see* **Note 15**).

4. Camera and software able to collect images at >12-bit depth (0-4095 intensity scale or greater) to accurately subtract ratios at 0.1% accuracies for reproducible FRET measurement.
5. If imaging CFP and YFP for FRET, a mercury, mercury/ xenon, or metal-halide light source is preferred (although not absolutely necessary) since they have a strong line in the 430–440 nm area for optimal excitation of the poorly detected CFP.
6. Image analysis software (*see* **Notes 16** and **17**) for accurate quantification of fluorescence amounts in a region of interest in the acceptor, donor, and FRET channels.

2.3 FRET Determination and Interpretation

1. Database software into which the quantified image data is imported for FRET calculation using simple maths. Microsoft Excel suffices for simple FRET determination.
2. Analytical software, such as GraphPad Prism, that conducts advanced statistics including nonlinear regression analysis to define curves that fit to the data (Fig. 3) and examination of distribution patterns (Fig. 2).
3. Calibration tools include cells expressing AR labeled only with the donor FP (CFP) to establish donor FP bleed through into the FRET channel and cells expressing acceptor (YFP)-only labeled AR to establish acceptor FP bleed through into the FRET channel (*see* **Notes 18** and **19**).
4. A set of expression vectors for calibration standards [46, 51]: These standards consist of proteins dual-tagged with the donor FP and acceptor FP positioned to provide variable levels of E [50]. Measurement of the standards on the operator's system is used to determine two parameters needed to conduct the measurements described in this chapter: (a) how well the energy transferred from the donor to the acceptor is detected as an increase in the FRET channel, relative to a decrease in the donor channel ("kfaD" in [51]) and (b) how well the same amount of donor FP and acceptor FP are detected in their specific channels ("kaD" in [51]).
5. A well-characterized, stable cell line for calibrating FP fluorescence intensities into protein concentrations: That calibration is required if one desires to compute biochemical kinetic parameters (Kd and Bmax) in "molar" terms in order to compare to values defined by traditional in vitro biochemical analyses. This extremely advanced application is described in detail elsewhere [55].

3 Methods

The methods and notes for transfecting cells in androgen-free media and measuring energy transfer were detailed in a prior chapter in the Methods series [41]. Those methods will only be summarized below to provide an overview of the method. New details about advanced procedures not discussed in the prior chapter are the focus of this chapter.

3.1 Image Capture

1. Transfections required for FRET analysis include "donor-only" and "acceptor-only" control cells (*see* **Notes 18** and **19**), which in this example are cells transfected with, respectively, CFP-tagged AR and YFP-tagged AR expression vectors. "Experimental" cells are transfected with CFP-AR-YFP (Figs. 1 and 2) or co-transfected with CFP-AR and a YFP-tagged target (Fig. 3) (*see* **Note 20**: transfections for intermolecular studies).
2. One day (or more) following transfection, add a ligand to a well. Imaging is conducted at specific times, depending on the question addressed, following ligand addition. *See* Ref. 41 for handling cover glasses for imaging.
3. Set integration (exposure) times and camera gain for "acceptor," "donor," and "FRET" channel collections (*see* **Note 21**: channel filter sets) necessary to average ~2000 intensity units on the 12-bit scale for the "dim" cells (*see* **Note 12**). When setting integration times for the entire experiment, keep in mind that, if FRET is higher under a different experimental condition, the donor channel intensity will decrease and the FRET channel intensity will increase. Avoid "image saturation" (*see* Ref. 41) in which the intensity readout is at the maximum in any pixel within any image. In a saturated image, actual image intensity will be greater than that quantified on the image, making it impossible to correct accurately for the FP bleed throughs, as required for FRET determination (discussed below).
4. Focus rapidly and collect all three channels rapidly and consistently from field to field, as if you were a robot. Discard image sets when you dwell on any field prior to collecting the images. Photobleaching of the FPs in such fields will change the FRET measurement (*see* **Note 22**) and introduce errors in your dataset.

3.2 Raw Fluorescence Quantification

1. Fluorescence quantification requires "background identification," defining areas in all channels where there are no cells or fluorescent debris; "background subtraction," removing the average background intensity; "segmentation," defining the cellular object in the image to be measured; and "quantification," determining fluorescence intensity values in the background-subtracted, segmented area for all three channels.
2. Background identification and subtraction is an error-prone component of FRET measurement. One can minimize the impact of background errors by increasing the cellular signal through higher expression of the FP-tagged factor. However, high expression is at odds with the desire to conduct the study under conditions in which trace levels of the FP-tagged factor interact with the cellular

environment in ways similar to the endogenous factor. Therefore, set collection conditions to image the low-intensity cells. Do not waste time trying to avoid high-intensity cells; saturated cells inadvertently captured will be instead flagged in the database and not analyzed.

3. A simple background identification and subtraction method is described below that is readily conducted by an inexperienced user on most image analysis platforms and is sufficient to provide the data quality of Figs. 2 and 3. After collecting the images, use your image analysis software to draw a box (a "region of interest") in a noncellular area beside the object you will measure. Transfer that background regions to all three of the channels, making sure that there is no "debris" in any of the channels that would introduce errors into the background subtraction. Most image analysis programs allow you to use this background region to create "background-subtracted images" in which the average fluorescence intensity within that background region is subtracted across the entire field for each of the three channels. Note that because the background generally is not uniform across the image (*see* **Notes 23** and **24**) this procedure will introduce background errors for cells far from the background box. So, for the simple procedure, it is best to repeat the procedure individually for each cell within a field placing the background box beside each specific cell (*see* **Note 25**). Ultimately, all background correction methods have some concerns (*see* **Note 26**) and the quality of the correction will be reflected in the overall quality of the data (*see* Fig. 3 as an example of good-quality data).
4. To define the object in which you will conduct your FRET measurement, use your image analysis software to draw a region of interest around the object. In our laboratory, we tend to use automated procedures that segment subcellular structures marked with other FPs (*see* **Note 27**).
5. Transfer the object's region of interest (ROI) to all three image channels (acceptor, donor, and FRET) in both the background-subtracted and original (not background-subtracted) images. Download the object data from all six image channels into your database, such as Excel. Data to download for FRET determination include object area, maximum intensity, and average intensity. Other quantifiers may be downloaded to, for example, compare the changes in FRET with the changes in distribution of the nuclear receptor or its cofactors following ligand addition [28, 54, 56].

3.3 Quality Control

1. With the downloaded data for each segmented ROI, discard datasets that contain any saturated pixels in any channel ("maximum" intensity in original images = 4095 using a 12-bit camera). This can be done readily in Excel using the "IF" function

to remove unqualified data or the "d" functions (such as dAverage) to include only qualified data in the average.

2. Background-subtracted intensity data from the cells expressing the bleed-through control vectors (*see* **item 3** of Subheading 2.3) provide measurements about how well the donor FP (or acceptor FP) appears in the FRET channel relative to the donor (or acceptor) channel. In the current example, the background-subtracted average intensity of the segment in donor FP-expressing cells measured in the FRET channel averaged 0.543 ± 0.017 that measured in the donor channel; for cells expressing the acceptor FP, the "bleed through" into the FRET channel was 0.140 ± 0.014 that measured in the acceptor channel. The bleed throughs of the donor FP and the acceptor FP into, respectively, the acceptor and donor channels were comparatively negligible (0.003 ± 0.004 and 0.015 ± 0.0130).
3. The bleed-through control calculations also provide an indicator of acceptor FP and donor FP average intensities required to obtain a consistently accurate measurement. For example, plotting the donor FP bleed through to the FRET channel (*y*-axis) against the donor intensity (*x*-axis) should provide a straight line at 0.543 across the range of donor intensities. If the 0.543 ratio measured starts to be measured less reproducibly at, for example, donor average intensities below 100 U (on the 0–4095 unit scale), then measurements below that area may be considered suspect. We typically flag such measurements in our database using Excel macros based upon the IF function.

3.4 FRET Determination

1. The "acceptor bleed-through-corrected FRET/donor" method represents a relatively simple way to determine the level of energy transfer (*see* **Note 28**). However, the corrected FRET/donor values will vary from instrument to instrument and even with each objective on an instrument, making them difficult to compare under different collection conditions [46].
2. A more advanced procedure enables calculation of the percentage of donor lost to energy transfer (E) from the bleed-through-corrected average intensities [46, 51, 57, 58]. Calibration standards and corrections for that procedure were described in the prior chapter [41]. Only a rudimentary discussion is provided here (*see* **Note 29**).

3.5 Determination of Molecularly Corrected Fluorescence Units

1. From the advanced calculations, one obtains (a) the amount of energy transfer (E, % donor lost to energy transfer); (b) the amount of acceptor FP measured (in acceptor channel fluorescence units); and (c) the amount of donor FP measured (in donor channel fluorescence units). However, if there is FRET, the amount of donor FP fluorescence measured is actually less than the amount of donor present since energy transferred to

the acceptor FP was not emitted as donor fluorescence. Knowing the percentage of donor FP fluorescence lost to energy transfer allows determination of the amount of donor actually present (in donor channel fluorescence) (*see* **Note 30**). By contrast, the amount of acceptor FP measured (excite acceptor FP only, measure acceptor FP in acceptor channel fluorescence units) is not affected by FRET and need not be corrected—do not confuse acceptor channel fluorescence with the acceptor fluorescence after donor excitation emitted within the FRET channel.

2. The number of fluorescence units in the donor and acceptor channels is not equivalent; for example, 500 U of fluorescence in the donor channel does not represent the same number of molecules as 500 U of fluorescence in the acceptor channel. However, the calibration standards, in which acceptor and donor FPs are positioned on the same carrier protein, allow also the determination of the relative donor and acceptor channel fluorescent units in the absence of energy transfer. This enables one to restate fluorescence units for both donor and acceptor in equivalent molecular terms (*see* **Note 31**).
3. In the interaction curve (methods discussed in the next section) shown in Fig. 3a, the amount of interaction (E, *y*-axis) is shown in relationship to the molecularly corrected fluorescence units of acceptor in cells which expressed an average of 916 equivalent units of donor FP-tagged AR. In Fig. 3, ~3000 equivalent fluorescence units of acceptor FP-tagged interacting protein were required to come close to maximal energy transfer (*see* **Note 32** and next section).

3.6 Curve Fitting for Biochemical Interactions

1. Curve fitting allows a more precise definition of the amount of acceptor FP-tagged interacting protein required to reach a specific degree of energy transfer. For each cell nucleus measured, we use a statistical program with nonlinear regression functions (such as GraphPad Prism) to compare the level of E measured (Fig. 3a, *y*-axis) against the amount of acceptor FP-labeled interacting factor in that nucleus (*x*-axis). That dataset is observed to fit well (*see* **Note 33**) to the equation $Y = \text{Bmax} \bullet X/(\text{Kd} \bullet X)$ that describes the interaction between two molecules when the concentration of one factor (the acceptor FP-labeled factor) is varied while keeping constant the donor FP-labeled interacting target.
2. The Bmax component of the equation is the maximal level of E in the presence of an infinitely large amount of acceptor FP-labeled factor. This "Emax" is technically not the Bmax (described in later comments) which is a biochemical parameter, but is rather a structural parameter [59]. Emax reflects the FP positions and orientations when all available (*see* **Note 34**)

donor FP-tagged factors are saturated with acceptor FP-tagged factor at equilibrium binding conditions (*see* **Note 35**). Note that the structure of the complex may be different at lower concentrations of acceptor FP-labeled interacting factor [55], which Emax provides no information about. We previously used mutants to lock particularly complexes into particular conformations [55] when confronted with deviations from the curve that implicated a changing structure as the level of the acceptor-labeled factor increased.

3. The amount of acceptor FP-labeled factor required to reach 50% of the Emax provides an indicator of interaction kinetics. The curve in Fig. 3a extrapolates that as 751 ± 81 U of acceptor FP labeled. If more acceptors were required, the interaction is of lower affinity. This value is an approximation of the equilibrium dissociation constant, Kd [55, 59], which is a standard parameter in biochemical measurement.
4. In classic in vitro biochemistry, the Kd should be calculated by plotting on the *x*-axis the concentration of "free" (i.e., "not bound in complex") interacting protein against the concentration of complex (*y*-axis). However, Fig. 3a plots the fluorescence units of all (i.e., bound and free) acceptor FP-labeled factor, which is inaccurate, against E which is not an equivalent concentration measurement. Figure 3a did provide the Emax (33.4% energy transfer for the 2-h incubation with DHT), from which can be estimated the concentration of free acceptor FP-labeled factor (proper *x*-axis) and the complex (proper *y*-axis) as follows.
5. For each data point (each nucleus in Fig. 3a), the proportion of donor-FP molecule bound can be described as E/Emax (*see* **Note 36**: assumptions) (*see* Table 1, column D; only a portion of the data is shown in Table 1). Since it is the donor FP that transfers energy, the intensity of donor fluorescence units (corrected for E, *see* **step 1** of Subheading 3.5) in the cell nucleus (column A) multiplied by the proportion of donor in the complex (E/Emax, column D) provides the donor fluorescence units in the complex (column E). If the complex is assumed to have a 1:1 stoichiometry of the donor FP-labeled and acceptor FP-labeled factors, then column E also represents the acceptor "equivalent" fluorescence units bound in the complex. That is subtracted from the total units of acceptor FP (converted into donor equivalent units) in the cell nucleus (column B) to provide the amount of "free" acceptor-labeled FP (column F, in equivalent fluorescence units). Excel or another spreadsheet program may be used to rapidly conduct these simple mathematical procedures, after which the resulting data in columns E (*y*-axis) and F (*x*-axis) are transferred to GraphPad Prism for nonlinear regression curve fitting according to the law of mass action.

Table 1
Calculation of the concentrations of complex (*y*-axis) and free acceptor (*x*-axis) within each cell nucleus

A	B	C	D	E	F
Donor[a]	Acceptor	E (%)	E/Emax (C/33.4)	[complex] (A•D)	[free Acc] (B – E)
902.3	94.3	2.87	0.0859	77.49	16.77
920.0	211.7	7.13	0.2135	196.45	15.21
936.4	604.5	15.57	0.4660	436.40	168.06
918.3	777.5	19.91	0.5961	547.43	230.06
920.0	856.7	16.83	0.5038	463.48	393.26
922.9	1715.4	26.59	0.7960	734.62	980.83
924.5	1974.1	30.89	0.9249	855.03	1119.03
934.9	2158.7	27.00	0.8083	755.67	1403.01

[a]The availability factor is multiplied by this column. Data shown assumes 100% availability. Data solution (**step 6** of Subheading 3.5) indicates that 58.6% of donor FP-labeled AR is available to interact

6. When conducting the above calculations for 2 h DHT incubation data points centered around 900 fluorescence units of E-corrected, donor FP-labeled AR (averaging 925 ± 144 U), the transformed data fit reasonably well to the equation *Y* = Bmax•X/(Kd•X) ($R^2 = 0.67$). However, the Bmax determined from this data was extrapolated as 795 fluorescence units of complex, whereas 925 fluorescence units of complex would be expected. As reported previously [55], the calculations in Table 1 assume, incorrectly, that all the donor FP-labeled AR in the cell is available to interact with acceptor FP-labeled factor. To examine what percentage of donor FP-labeled AR might be available to interact, replace column A with the amounts of donor FP-labeled AR multiplied by an availability factor; for example, on the first row in Table 1, 50% of 902.3 fluorescence units (column A) would become 451.1 donor FP-labeled factor available to interact. Create, in the database, a matrix of availability assumption for the entire data set (*see* **Note 37**: assumptions) and examine the resulting columns E and F for fitting to the curve in GraphPad Prism. In the current example, 45.5% availability provided a solution (Fig. 3b) for which the Bmax determined from that curve (421 ± 26) agreed with the average 925 donor FP-tagged AR fluorescence units (i.e., 0.455 × 925 = 421) and that fit well to the curve ($R^2 = 0.92$, normally distributed, no aberrant runs). The Kd in this example showed that 505 ± 97 fluorescence units of free acceptor FP-labeled factor are present when half

of the available donor FP-labeled AR ($0.5 \times Bmax = 210$ U) is bound in a complex with acceptor FP-labeled factor, presumably in a 1:1 stoichiometry with 210 unit of bound acceptor FP-labeled AR.

7. With the above calculations, it remains impossible to compare the affinity (reflected by the Kd) determined in fluorescence units to the Kd determined in vitro (in molar concentrations). It also would be impossible to compare results in fluorescence units collected in different laboratories, or even within the same laboratory on a different instrument. However, if one possesses a cell line stably expressing, at a uniform level in most cells, a factor fused to one of the FPs used in the study, then one can conduct studies to determine the average concentration of the FP-labeled factor in each cell and compare that to the fluorescence units [55]. After that, one simply exchanges the fluorescence units in Fig. 3b with the molar concentrations to obtain true biochemical values (*see* **Note 38**).

3.7 Pixel-by-Pixel Analysis

1. The size of the region of interest can be as small as a pixel which, as described in Fig. 2b, can be informative. The data from each pixel can be downloaded into a database with the calculations performed on each pixel. Reconstructing the image would require that each pixel is associated with a position in space for reconstruction. Some image analysis programs present in most laboratories may not have this capability. However, those programs likely do have the ability to perform arithmetic on images (*see* **Note 39**). One should be aware of errors that can be introduced by image calculations for pixels with no energy transfer (*see* **Note 40**).
2. With a FRET image calculated, numeric data still needs to be extracted in order to create frequency distribution graphs like those shown in Fig. 2. This is achieved by sequential thresholding the images within the object region of interest (e.g., threshold pixels with 0% energy transfer and download data; then from >0 to 1.0% energy transfer and download data; >1 to 2.0% energy transfer). That information is created by using an automated macro in the image analysis software (called a journal in the Metamorph software program we often use), with the data exported into Excel at each step. For the cellular biochemistry analysis (Fig. 3), sequentially threshold on pixels with progressively higher acceptor fluorescence intensities, transfer those pixels to the donor and E channels, and export that data into Excel.
3. With the image collection of all three channels occurring at different times, and with the rapid movement of the AR with time [33, 38, 60], be aware that the pixel-by-pixel measurements are being conducted on molecules streaming through an area, not on single molecules that remain locked in a single pixel.

4 Notes

1. For example, energy excited within a CFP donor will transfer to a YFP acceptor 50% of the time when the centers of the energy fields are 49.2 Å apart but only 5% when they are 80 Å apart [61].
2. When two different FP-tagged proteins are present within a membrane, the FPs are constrained in two-dimensional space, not three-dimensional space, and their effective concentrations may become high enough for FRET to occur even in the absence of an interaction between the proteins to which the FPs are attached [62]. In the three-dimensional space, donor and acceptor FPs attached to two different proteins will seldom be sufficiently concentrated unless those two proteins interact and bring the FPs close to each other.
3. The orientation factor is an unknown in most FRET applications [63–65]. If the FPs are freely rotating, any experimentally induced alterations in E reflect primarily a change in FP distance. One can partially check the assumption of no orientation constraint by exchanging the donor and acceptor FPs at the same positions. If E remains the same, then the changes in energy transfer are less likely to have an orientation component.
4. It is impractical to follow the same cells over a 24-h time period unless one has access to automated systems with cell tracking capabilities. Unless there are experiment reasons for repeated measurements on the same cells, we more typically add ligand to different wells, then capture images of one well at one time point, and then discard the cells. The lack of repeat measures on the same cell also eliminates the photobleaching errors introduced by repeated illumination. Photobleaching changes FRET (e.g., reducing the level of intact acceptor will reduce energy transfer) and accurate FRET depends on rapid, often automated [66], capture methods that minimize photobleaching.
5. It can take 1–2 s to acquire the three images required to calculate FRET. Given the rapid dynamics of nuclear receptors within the cell, the pixel-by-pixel calculations should not be considered to be a measurement of individual ARs at individual sites. Rather, FRET will provide information about the average conformations of a number of ARs transiting through each pixel during image capture.
6. Pixel-by-pixel analysis can provide some surprising information. We have seen instances in which the addition of a ligand did not change the average FRET level but instead changed how "tight" FRET was throughout the cell (e.g., $20 \pm 10\%$ E to $20 \pm 5\%$ E). This may reflect the ability of a ligand to "lock" a flexible receptor into a preferred conformation throughout the cell.

7. Although it is common to consider FRET as originating from the dimerization of the FP-tagged ARs, the presence of donor and acceptor FPs in close enough proximity within any complex of any stoichiometry will lead to FRET.
8. Objectives with high numerical apertures (NA) improve the detection of fluorescence from FPs expressed at low levels (*see* **Note 12**). High NA objectives generally have short working distances that enable imaging only through cover glass-thick materials. In our laboratory, we more commonly grow cells in thin-bottomed 384-well plates, such as those available from Matrical or from Greiner Bio-One, and image on an inverted fluorescence microscope.
9. Select FPs with a high degree of overlap of the donor emission with the acceptor excitation (Fig. 1e), but that still can be distinguished in the donor and acceptor collection channels. The donor FP is preferred to have a high quantum yield (a high likelihood of emitting the absorbed energy) and the acceptor is desired to be excited relatively well (high molar extinction coefficient). The CFP and YFP pair used here is inferior to next-generation FPs discussed elsewhere [67].
10. All cellular methods rely on the creation of FP-tagged nuclear receptors or cofactors that retain their functional properties. The first step is to compare the FP-fused factor with the unfused factor in a battery of functional assays usually conducted by transfecting or infecting cells with expression vectors for the factors. This establishes the subset of functional activities that the FP-tagged factor remains competent in regulating. One must limit the interpretation of the information received from the fluorescence methods to molecular activities that are retained following FP fusion. This is not a constraint only for FRET measurement, but is true for any experiment in which the protein studied is modified in any way, be it by mutation, by truncation or deletion, or by addition of an epitope or domain (most FPs are ~27 kD).
11. Cell culture media, serum, some tissue culture plates or coatings, and even some transfection reagents contain fluorescent molecules. Plate cells in different media or plates and image under your collection conditions to define collection conditions with the lowest intrinsic fluorescence in which your cell type remains healthy. Tissue culture-treated multi-well plates from Greiner Bio-One and RPMI 1640 (of the media obtained through our institutional cell culture facility) are optimal for the cells in this study. Lower background fluorescence generally is also achieved when cells are grown on glass cover slips or on glass-bottom plates, provided that glass is suitable for that cell type. Regardless, there will always be some background, which must be corrected for accurate FRET determination.

12. Test different transfection reagents by packaging a single FP-labeled expression vector. Some reagents (such as lipofectamine in our hands) show less cell-to-cell fluctuation in fluorescence levels. When capturing images, avoid the natural tendency to set the exposure times based on the brightest cells—there may be 5- to 20-fold more cells of a lower intensity. The intensely bright cells often are those in which the FP-tagged protein is expressed at levels far exceeding that which an endogenous factor would normally be found. Using such high-expressing cells therefore would negate the rationale for investigating the protein within the context of the competitive and cooperative influences present with the cell.
13. Select "donor" excitation and emission filters (*see* **Note 21**) that detect only the donor FP with no zero emissions evident when the acceptor FP is expressed in the absence of donor FP. Similarly, select filters for a "clean" detection of the acceptor FP only in the "acceptor" channel. The "FRET" channel will contain emissions from both the donor and the acceptor FP, which must be precisely corrected for in the FRET method.
14. Excitation and emission filters are best accommodated in filter wheels (e.g., Sutter Instrument Company) and used together with a single, multi-bandpass dichroic mirror. The filter wheel/single dichroic configuration minimizes pixel shifts that typically occur when using different filter cubes and it enables automated, rapid image collection that minimizes cell movements during the collection of the three channels needed for FRET determination.
15. FRET measurements depend on corrections to accurately quantify fluorescence values at each pixel. The correction factors generally are applied to all pixels, regardless of intensity. That is valid because the physical properties of each FP do not vary with FP amount. However, many cameras (and photomultiplier tubes for confocal collections) are not quantitatively linear, resulting in corrections that vary with intensity. To test for camera linearity prior to purchase, capture the same images at different exposure times (e.g., 400 and 100 ms) and ensure that the background-subtracted fluorescence intensity measured is indeed one-quarter at 100 ms compared to 400 ms. In our experience, you should expect a 0.250 ± 0.002 intensity ratio for a fourfold difference in exposure times across all intensity levels [51]. If your camera is not linear, FRET analysis programs have been developed in which the ratios are determined across the dynamic range of measurements and subtracted [57, 58].
16. Publicly available freeware such as ImageJ can be used. We use MetaMorph software (Molecular Devices) as we tend to conduct more detailed, automated procedures (often in 384-well format) that employs more advanced procedures for defining objects and backgrounds [66] than is discussed here.

17. Many commercially available software programs come preloaded with FRET packages that are often based on erroneous understandings of FRET, or that provide alternative methods from which many users can make ill-informed selections. It is best to understand the concepts outlined here and to evaluate if those packages comply with accurate procedures.
18. Because of overlap in their excitation spectra, both CFP and YFP will be excited by the excitation light used for "CFP" excitation. Thus YFP emits light in the "FRET" channel. CFP also emits light in the "FRET" channel as many CFP-emitted photons are not "cyan" at all, being photons of green, yellow, and even red light. Thus, when both FPs are present, the resulting FRET channel image is not representative of just FRET, but a mixture of light originating independently from the donor FP and from the acceptor FP, often referred to as "bleed throughs," and a small amount of genuine energy transfer buried among those much larger signals. The control cells (expressing only the CFP-labeled factor or the YFP-labeled factor) are used to determine the "bleed throughs" of the donor FP and the acceptor FP into the FRET channels, which is detailed in the prior chapter in this series [41]. The fundamental consideration is that the operator rarely is able to simply look down the eyepiece of the microscope and visually determine that there is energy transfer.
19. In theory, it is possible to use FPs in which the contributions of the individual donor and acceptor FPs to the FRET channel are so minimal that FRET could be directly viewed. However, under such conditions, there also is very little overlap of the donor emission energy with the energy required for acceptor excitation. These conditions thus reduce the amount of FRET to be measured. Overall, stronger overlap results in stronger, more readily quantifiable energy transfers that are buried in stronger also more readily quantified bleed throughs.
20. Interpreting protein–protein interactions relies on transfection conditions in which each cell expresses a relatively constant level of donor FP-labeled factor but a wide divergence in acceptor-labeled factor expression (Fig. 3). This is achieved in part by engineering the vector so that the donor-labeled factor is expressed weakly compared to the acceptor-labeled factor. Thus, in a standard study in our lab in which 1000 ng of all vectors is to be transfected, 900 ng will be of a weak expression vector for the donor-labeled factor. To get the nice "spread" in acceptor-labeled expression from the remaining 100 ng, we typically prepare three transfections in which the 900 ng of donor vector is mixed with one of (a) 100 ng of the expression vector for the acceptor-labeled factor; (b) 30 ng of the acceptor vector and 70 ng of "blank" vector with no protein or FP

cDNA inserted, and (c) 10 ng acceptor and 90 ng blank vector. After the three conditions are incubated with the transfection reagent, they are mixed together and added to the cells.

21. A typical set of filter combinations for CFP and YFP FRET include (a) excite with 496–505 nm/collect 520–550 nm emissions (collects YFP with no collection of CFP); (b) excite with 431–440 nm/collect 455–485 nm emissions (collects CFP with no collection of YFP); and (c) excite with 431–440 nm/collect 520–550 nm emissions (collects CFP and YFP "bleed throughs" into this channel plus any energy transferred from CFP to YFP).

22. When photobleached, the acceptor FP is no longer available for the donor FP to transfer energy to. If there is any photobleaching, the amount of FRET measured is less than what had been present. In fact, acceptor photobleaching is the basis of a FRET measurement method for determining E from the amount of donor FP fluorescence gained after photobleaching [5, 68]. This is a very simple alternative method for novice FRET practitioners although one should be concerned about any simultaneous bleaching of the donor that would impact FRET calculation.

23. The background has multiple components: (a) fluorescent molecules in the media which can vary at different parts of the field because of non-even illumination by the excitation light; (b) ambient light scattering throughout the field; and (c) the low-level intensity from the camera itself (camera noise). "Flat-field correction" procedures are available to correct for that non-even illumination [41] but, for the simple procedure described here, it is sufficient to place the background box in an area adjacent to the cell which closely approximates the background within the cell.

24. Our laboratory tends to use sophisticated, but cumbersome, methods that flat-field correct and accurately remove background [41]. Those procedures are accurate enough to enable FRET determination even under extremely low signal-to-noise conditions desired for examining FP-tagged factors expressed at "tracer" levels in the cell (*see* **Note 12**).

25. It is not necessary to create "background-subtracted images." One can simply place a background region beside each cell in which each segmentation region is placed, and then transfer that value to the database. The subtraction then is done within the database (such as Excel) to which the fluorescence data is transferred.

26. There are unknowns in any background-correction procedure that may introduce errors. If your optical section captures fluorescence entirely within the cell (no media), then subtracting

the media component of the background fluorescence may be inappropriate. More typically, your images will likely contain a mixture of cells at different depths and with variable media components. In short, one can minimize but never completely eliminate the errors associated with determining background.

27. In our high-throughput studies, we use cell lines expressing a nuclear marker tagged with a red fluorescent protein [66, 69]. The nuclear marker requires the collection of a fourth fluorescence channel but enables automated segmentation of the nuclei that is transferred to all images. AR-YFP fluorescence in the AR-specific channel is used to also define the margins of the cell associated with each nucleus. This enables quantification of fluorescence values in both the nucleus and cytoplasmic compartments. With that information, it is possible to compare FRET measurements of AR structure and biochemistry with parallel measurements of total AR–FP level, its distribution between the nuclear and cytoplasmic compartments, and the variations in its distributions within those compartments.
28. For the FRET/donor approach, the acceptor bleed-through calibration first is used to calculate and subtract the acceptor FP bleed-through intensities present in the FRET and donor channels. For an example of one cell in which the background-subtracted intensities in the acceptor, donor, and FRET channels were 1171, 818, and 798, the acceptor bleed throughs into the donor and FRET channels are, respectively, 17 and 164 (1.5 and 14.0% of 1171). The ratio of the bleed-through-subtracted intensity in the FRET channel ($634 = 798 - 164$) over the bleed-through-subtracted intensity in the donor channel ($801 = 818 - 17$) is 0.792, which is greater than that which would have come from the donor FP alone in the absence of FRET (0.543: the FRET/donor channel intensity ratio from the control cells expressing only donor FP). The ratio of 0.792 indicates energy transfer, which causes an elevation in the FRET channel intensity and a diminution in the donor channel intensity energy transfer (*see* also Ref. 41). This is a simple calculation that can be used within a laboratory without the need for further calibrations.
29. Calibration standards are used to determine a constant (referred to as KfaD in [51]) that converts FRET/donor into E [46, 51]. Like the bleed-through values, the constant is affected by the filter set, camera, and objectives used on a specific instrument. In the current example, instrument calibrations established KfaD as 1.407. With a FRET/donor ratio of 0.792 within a cell nucleus, E is calculated (*see* Ref. 51) as $(0.792 - 0.543)/(1.407 + 0.792 - 0.543)$ or 15.0%. Note that a doubling of the FRET/donor ratio does not lead to a doubling of E. Thus, only E can be applied for the more sophisticated calculations described in this chapter.

30. In the example provided in **Note 29**, E was measured as 15% and the donor fluorescence was measured as of 801 fluorescence units. The amount of donor actually present would be equivalent to 801/(1 − 0.15) or 942 U of donor channel fluorescence.
31. The instrument calibration determined a constant, KaD, which describes 1000 U of donor FP fluorescence in the donor channel as equivalent to 1229 fluorescence units of the same number of acceptor FP molecules emitting in the acceptor channel. In the example provided (**Notes 28** and **29**), the 942 U of donor E-corrected CFP fluorescence was measured for a CFP–AR–YFP probe that contains an equimolar amount of YFP. 1171 U of YFP fluorescence was measured in the same cell, which would be equivalent to 953 (1171/1.229) equivalent units of CFP fluorescence. The deviation from 942 reflects the cell-to-cell measurement errors that are sometime less than, and sometime greater than, the measured amounts. Be aware of those 1% measurement errors when interpreting your FRET measurements with a statistically insufficient number of cells.
32. The current example is for the oligomerization of the AR. Therefore, if establishing how much acceptor-FP-labeled AR is required to reach 50% saturation of the binding of the donor P-labeled AR, one also has to consider that there is a competition between the binding of the donor FP-labeled AR with itself that also changes as the level of acceptor FP-labeled AR rises with a relatively constant level of donor FP-labeled AR. That correction has not been applied in the current example which is focused on providing readers with the basic methodology.
33. Establishing the degree to which the data points fit to a curve was detailed previously [55]. The major parameters to be established include (a) the "goodness of fit" (R^2), which in the Fig. 3a example was 0.87 (1 is maximal); (b) whether the data is distributed on average normally around the best-fitting curve (yes in Fig. 3a); and (c) whether there are any "runs" of data points that appear consecutively above or below the best-fitting curve in a pattern that appears nonrandom (no in Fig. 3a). Such "runs analyses" can be informative about elements such as a change in the stoichiometry of the complex that forms at low and high levels of SRC cofactor interacting with ERα-interacting factors [55].
34. This interaction curve is classically fit to data points collected by measuring the amounts of complex formed in vitro in the presence of increasing amounts of one purified factor while holding the target protein at a constant concentration. Bmax is assumed to occur when 100% of the target protein is interacting. However, in the cell, not all of the target protein (which

must be labeled with the donor FP) is available to interact with acceptor protein (*see* Subheading 1). This represents a primary distinction between the "cellular biochemistry" method described here and classic in vitro biochemistry approach.

35. Equilibrium binding assumes that sufficient time for interaction has been provided to reach equilibrium (2 h after DHT addition for Fig. 3). The curve fit for data collected 20 min after DHT addition was poorer ($R^2 = 0.61$; non-normal instead of 0.87; normal at 2 h) possibly reflecting the different lag times in DHT response within individual cells. The overall Emax at 20 min of DHT response also showed more error (33.8 ± 7.3%) than at 2 h (33.4 ± 1.7%), possibly reflecting the failure to reach equilibrium response by 20 min.
36. Assumptions in these calculations discussed previously [55] include the following: (a) no change in the structure of the complex, which affects E, at different levels of acceptor FP-labeled factor present; (b) the fluorescence intensities collected on a two-dimensional image reflect molecular concentrations in a three-dimensional volume.
37. The availability factor includes an assumption about a 1:1 stoichiometry. For example, if the stoichiometry is 2:1 of donor-labeled factor to acceptor-labeled factor, then "availability" would be 50% at most. Indeed, in the original publication describing the method, the assumption of a 1:1 stoichiometry was invalid for certain interactions of the estrogen receptor (alpha isoform) with three different cofactors [55]. Other factors that may reduce the availability were described in Subheading 1 here, as well as in [55].
38. In vitro, the Bmax will vary with how much of the "constantly held" factor that is put into the study. However, in the cell, the Bmax can represent a ceiling that is limited by the requirement for other factors within the cell as was observed for SRC interaction with estrogen receptor-alpha [55]. Thus, in addition to providing details about affinities (Kd) and structure (Emax) of the complex within the cell, the cellular biochemistry methods also provide information about cellular limitations on complex formation (Bmax and availability). In the AR oligomerization study describe in this chapter, Bmax increased with increasing donor level (data not shown) suggesting that if any cellular factors limit AR oligomerization, they are present in the cell at concentrations higher than the expression levels we achieved for the probes in this study.
39. The example below is provided for conducting image-based arithmetic using the Metamorph image analysis program starting with the background-subtracted acceptor, donor, and FRET images (*see* **Notes 23–28**). Create an image of the acceptor FP bleed through to the FRET channel by multiplying the

acceptor channel image (i.e., all pixels within the image) by 140/1000 (14.0% bleed through). Subtract that image from the FRET channel image. Similarly subtract the 1.5% bleed through of the acceptor FP into the donor channel. To obtain a FRET/donor image, divide the acceptor bleed-through-corrected FRET channel image by the acceptor bleed-through-corrected image of the donor channel. Because this ratio is a fraction (example: 0.792 in **Note 28**) and because the images only display whole numbers, it is necessary to multiple the bleed-through-corrected FRET channel image (the numerator) by 1000 prior to creating the FRET/donor image. For calculating E (*see* **Note 29**), this FRET/donor image first is used to subtract the FRET/donor constant of the donor FP only (0.543, or 543 since all has been multiplied by 1000). The resulting image is the numerator used for E determination; the denominator image is created by adding KfaD to the numerator image (1407, after multiplying 1.407 by 1000). Those two images are then divided by each other but, since this results in a fraction (% donor lost to energy transfer), the numerator must again be multiplied by 1000 prior to division. Thus, a pixel of "100" will represent 10% energy transfer.

40. When performing arithmetic on images as in **Note 39**, be aware that negative numbers are scored as "0." A region of interest that scores as zero energy transfer when calculating throughout the region will consist of negative and positive pixels through measurement error. Thus, when averaging the pixels calculated through image subtraction, this will appear as a small number of pixels with positive values and a lot with negative values.

References

1. Tsien RY (1998) The green fluorescent protein. Annu Rev Biochem 67:509–544
2. Zhang J, Campbell RE, Ting AY et al (2002) Creating new fluorescent probes for cell biology. Nat Rev Mol Cell Biol 3:906–918
3. Tsien RY (2003) Imaging imaging's future. Nat Rev Mol Cell Biol (Suppl):SS16–SS21
4. Chudakov DM, Lukyanov S, Lukyanov KA (2005) Fluorescent proteins as a toolkit for in vivo imaging. Trends Biotechnol 23:605–613
5. Day RN, Schaufele F (2005) Imaging molecular interactions in living cells. Mol Endocrinol 19:1675–1686
6. Giepmans BN, Adams SR, Ellisman MH et al (2006) The fluorescent toolbox for assessing protein location and function. Science 312:217–224
7. Ishikawa-Ankerhold HC, Ankerhold R, Drummen GP (2012) Advanced fluorescence microscopy techniques – FRAP, FLIP, FLAP, FRET and FLIM. Molecules 17:4047–4132
8. Matsuda T, Nagai T (2014) Quantitative measurement of intracellular protein dynamics using photobleaching or photoactivation of fluorescent proteins. Microscopy (Oxf) 63:403–408
9. Stasevich TJ, Mueller F, Michelman-Ribeiro A et al (2010) Cross-validating FRAP and FCS to quantify the impact of photobleaching on in vivo binding estimates. Biophys J 99:3093–3101
10. Digman MA, Gratton E (2012) Scanning image correlation spectroscopy. Bioessays 34:377–385
11. Ries J, Schwille P (2012) Fluorescence correlation spectroscopy. Bioessays 34:361–368
12. Sprague BL, McNally JG (2005) FRAP analysis of binding: proper and fitting. Trends Cell Biol 15:84–91
13. Carrero G, McDonald D, Crawford E et al (2003) Using FRAP and mathematical modeling to determine the in vivo kinetics of nuclear proteins. Methods 29:14–28

14. Kerppola TK (2006) Design and implementation of bimolecular fluorescence complementation (BiFC) assays for the visualization of protein interactions in living cells. Nat Protoc 1:1278–1286
15. Day RN, Periasamy A, Schaufele F (2001) Fluorescence resonance energy transfer microscopy of localized protein interactions in the living cell nucleus. Methods 25:4–18
16. Voss TC, Demarco IA, Day RN (2005) Quantitative imaging of protein interactions in the cell nucleus. Biotechniques 38:413–424
17. Vogel S, Thaler C, Koushik SV (2006) Fanciful FRET. Sci STKE 2006(331):ra2
18. Piston DW, Kremers GJ (2007) Fluorescent protein FRET: the good, the bad and the ugly. Trends Biochem Sci 32:407–441
19. Sun Y, Rombola C, Jyothikumar V et al (2013) Förster resonance energy transfer microscopy and spectroscopy for localizing protein-protein interactions in living cells. Cytometry A 83:780–793
20. Demarco IA, Periasamy A, Booker CF et al (2006) Monitoring dynamic protein interactions with photoquenching FRET. Nat Methods 3:519–524
21. Nguyen TA, Sarkar P, Veetil JV et al (2012) Fluorescence polarization and fluctuation analysis monitors subunit proximity, stoichiometry, and protein complex hydrodynamics. PLoS One 7:e38209
22. Hager GL (1999) Studying nuclear receptors with green fluorescent protein fusions. Methods Enzymol 302:73–84
23. McNally JG, Müller WG, Walker D et al (2000) The glucocorticoid receptor: rapid exchange with regulatory sites in living cells. Science 287:1262–1265
24. Becker M, Baumann C, John S et al (2002) Dynamic behavior of transcription factors on a natural promoter in living cells. EMBO Rep 3:1188–1194
25. Weatherman R, Chang C, Clegg N et al (2002) Ligand-selective interactions of ER detected in living cells by fluorescence resonance energy transfer. Mol Endocrinol 16:487–496
26. Nagaich AK, Rayasam GV, Martinez ED et al (2004) Subnuclear trafficking and gene targeting by steroid receptors. Ann N Y Acad Sci 1024:213–220
27. Hinojos CA, Sharp ZD, Mancini MA (2005) Molecular dynamics and nuclear receptor function. Trends Endocrinol Metab 16:12–18
28. Schaufele F, Carbonell X, Guerbadot M et al (2005) The structural basis of androgen receptor activation: intramolecular and intermolecular amino-carboxy interactions. Proc Natl Acad Sci U S A 102:9802–9807
29. De S, Macara IG, Lannigan DA (2005) Novel biosensors for the detection of estrogen receptor ligands. J Steroid Biochem Mol Biol 96:235–244
30. Koterba KL, Rowan BG (2006) Measuring ligand-dependent and ligand-independent interactions between nuclear receptors and associated proteins using Bioluminescence Resonance Energy Transfer (BRET). Nucl Recept Signal 4:e021
31. Martinez ED, Hager GL (2006) Development of assays for nuclear receptor modulators using fluorescently tagged proteins. Methods Enzymol 414:37–50
32. Griekspoor A, Zwart W, Neefjes J et al (2007) Visualizing the action of steroid hormone receptors in living cells. Nucl Recept Signal 5:e003
33. Klokk T, Kurys P, Elbi C et al (2007) Ligand-specific dynamics of the androgen receptor at its response element in living cells. Mol Cell Biol 27:1823–1843
34. Padron A, Li L, Kofoed E et al (2007) Ligand-selective interdomain conformations of estrogen receptor-alpha. Mol Endocrinol 21:49–61
35. Paruthiyil S, Cvoro A, Zhao X et al (2009) Drug and cell type-specific regulation of genes with different classes of estrogen receptor beta-selective agonists. PLoS One 4:e6271
36. Jones JO, An WF, Diamond MI (2009) AR Inhibitors identified by high-throughput microscopy detection of conformational change and subcellular localization. ACS Chem Biol 4:199–208
37. Mueller F, Mazza D, Stasevich TJ et al (2010) FRAP and kinetic modeling in the analysis of nuclear protein dynamics: what do we really know? Curr Opin Cell Biol 22:403–411
38. Kirli HZ, Saatcioglu F (2011) Methods to study dynamic interaction of androgen receptor with chromatin in living cells. Methods Mol Biol 776:131–145
39. van Royen ME, Dinant C, Farla P et al (2009) FRAP and FRET methods to study nuclear receptors in living cells. Methods Mol Biol 505:69–96
40. Nishi M (2010) Imaging of transcription factor trafficking in living cells: lessons from corticosteroid receptor dynamics. Methods Mol Biol 647:199–212
41. Schaufele F (2011) FRET analysis of androgen receptor structure and biochemistry in living cells. Methods Mol Biol 776:147–166
42. Sun Y, Day RN, Periasamy A (2011) Investigating protein-protein interactions in living cells using fluorescence lifetime imaging microscopy. Nat Protoc 6:1324–1340
43. Sun Y, Hays N, Periasamy A et al (2012) Monitoring protein interactions in living cells

with fluorescence lifetime imaging microscopy. Methods Enzymol 504:371–391
44. Piston DW, Rizzo MA (2008) FRET by fluorescence polarization microscopy. Methods Cell Biol 85:415–430
45. Levitt JA, Matthews DR, Ameer-Beg SM et al (2009) Fluorescence lifetime and polarization-resolved imaging in cell biology. Curr Opin Biotechnol 20:28–36
46. Koushik SV, Chen H, Thaler C et al (2006) Cerulean, Venus, and VenusY67C FRET reference standards. Biophys J 91:L99–L101
47. Pelet S, Previte MJ, So PT (2006) Comparing the quantification of Forster resonance energy transfer measurement accuracies based on intensity, spectral, and lifetime imaging. J Biomed Opt 11(34017):1–11
48. Förster T (1948) Zwischenmolekulare energiewanderung und fluoreszenz. Ann Phys 6: 54–75
49. Förster T (1959) Transfer mechanisms of electronic excitation. Discuss Faraday Soc 27:1–17
50. Chen H, Puhl HL, Koushik SV et al (2006) Measurement of FRET efficiency and ratio of donor to acceptor concentration in living cells. Biophys J 91:L39–L41
51. Kofoed EM, Guerbadot M, Schaufele F (2008) Dimerization between aequorea fluorescent proteins does not affect interaction between tagged estrogen receptors in living cells. J Biomed Opt 13(031207):1–15
52. Stryer L, Haugland RP (1967) Energy transfer: a spectroscopic ruler. Proc Natl Acad Sci U S A 58:719–726
53. He B, Kemppainen JA, Wilson EM (2000) FXXLF and WXXLF sequences mediate the NH2-terminal interaction with the ligand binding domain of the androgen receptor. J Biol Chem 275:22986–22994
54. van Royen ME, van de Wijngaart DJ, Cunha SM et al (2013) A multi-parameter imaging assay identifies different stages of ligand-induced androgen receptor activation. Cytometry A 83:806–817
55. Kofoed EM, Guerbadot M, Schaufele F (2010) Structure, affinity, and availability of estrogen receptor complexes in the cellular environment. J Biol Chem 285:2428–2437
56. van Royen ME, Cunha SM, Brink MC et al (2007) Compartmentalization of androgen receptor protein-protein interactions in living cells. J Cell Biol 177:63–72
57. Elangovan M, Wallrabe H, Chen Y et al (2003) Characterization of one- and two-photon excitation fluorescence resonance energy transfer microscopy. Methods 29:58–73
58. Wallrabe H, Periasamy A (2005) Imaging protein molecules using FRET and FLIM microscopy. Curr Opin Biotechnol 16:19–27
59. Chen H, Puhl HL, Ikeda SR (2007) Estimating protein-protein interaction affinity in living cells using quantitative Förster resonance energy transfer measurements. J Biomed Opt 12(054011):1–9
60. Marcelli M, Stenoien DL, Szafran AT et al (2006) Quantifying effects of ligands on androgen receptor nuclear translocation, intranuclear dynamics, and solubility. J Cell Biochem 98:770–788
61. Patterson GH, Piston DW, Barisas BG (2000) Förster distances between green fluorescent protein pairs. Anal Biochem 284:438–440
62. Kiskowski MA, Kenworthy AK (2007) In silico characterization of resonance energy transfer for disk-shaped membrane domains. Biophys J 92:3040–3051
63. van der Meer BW (2002) Kappa-squared: from nuisance to new sense. J Biotechnol 82: 181–196
64. Vogel SS, van der Meer BW, Blank PS (2014) Estimating the distance separating fluorescent protein FRET pairs. Methods 66: 131–138
65. Ivanov V, Li M, Mizuuchi K (2009) Impact of emission anisotropy on fluorescence spectroscopy and FRET distance measurements. Biophys J 97:922–929
66. Schaufele F (2014) Maximizing the quantitative accuracy and reproducibility of Förster resonance energy transfer measurement for screening by high throughput widefield microscopy. Methods 66:188–199
67. Day RN, Davidson MW (2012) Fluorescent proteins for FRET microscopy: monitoring protein interactions in living cells. Bioessays 34:341–350
68. Bastiaens PI, Majoul IV, Verveer PJ et al (1996) Imaging the intracellular trafficking and state of the AB5 quaternary structure of cholera toxin. EMBO J 15:4246–4253
69. Krylova I, Kumar RR, Kofoed EM et al (2013) A versatile, bar-coded nuclear marker/reporter for live cell fluorescent and multiplexed high content imaging. PLoS One 8:e63286

Chapter 7

Posttranslational Modifications of Steroid Receptors: Phosphorylation

Dagmara McGuinness and Iain J. McEwan

Abstract

The detection of phosphorylation status of proteins has become a critical component of the analysis of activity, localization, and turnover studies of most proteins, particularly for those involved in signaling. The androgen receptor is no exception to this rule with its localization, transcriptional activity, and interactions determined by a series of key phosphorylations on serine residues. Here we have presented a series of techniques for the investigation of the phosphorylation status and intracellular localization of the androgen receptor after hormone and growth factor stimulation of cells in culture (in vitro) and in prostate cancer tissue (in vivo). Modified methods for immunohistochemistry, immunoblotting and immunofluorescence detection with high efficacy for the measurement and monitoring of androgen receptor are presented here alongside examples of their use.

Key words Androgen receptor, Phosphorylation, Immunoprecipitation, Immunohistochemistry, Immunofluorescence, Western blot

1 Introduction

The detection of phosphorylated proteins has become a critical research technique in determining not only the presence of a particular protein of interest, but also for determining its activity. The presence or absence of a specific phosphorylation modification can activate, deactivate, or modulate the activity of a given protein [1]. Nuclear receptors are well recognized as phospho-proteins and recent years have seen considerable progress in the identification and characterization of phosphorylation sites and receptor function [2, 3]. The androgen receptor (AR; NR3C4) is a member of the steroid receptor subfamily and mediates the action of the androgens, testosterone and dihydrotestosterone, within cells [3]. This protein is critical for the development of the male sex tissues, including the prostate gland [2]. In fact, aberrant posttranslational modification of AR has been linked to disease, including cancer [4]. There are a number of phosphorylation sites in the AR protein,

Iain J. McEwan (ed.), *The Nuclear Receptor Superfamily: Methods and Protocols*, Methods in Molecular Biology, vol. 1443, DOI 10.1007/978-1-4939-3724-0_7, © Springer Science+Business Media New York 2016

which have been associated with specific functions, and considerable effort has led to the identification of kinases and phosphatases responsible for modifying the receptor protein. In the amino-terminal domain (NTD), serine 81 was found to be phosphorylated by cyclin-dependent kinases (CDKs) and is phosphorylated in response to androgens [5], and is involved in the transactivation of AR [6, 7]. Serines 213 and 791 are targeted by the AKT/PI3K pathway and are generally considered to reduce AR activity [8–10], while serine 293, also within the AR-NTD, can be phosphorylated by Aurora-A, potentially linking it to poor prognosis in prostate cancer [11]. Phosphorylation of serine 308 modulates transcriptional activity [5]. Serine 515 is phosphorylated in response to epidermal growth factor (EGF), and this response is linked to phosphorylation at serine 578 [12]; modification of serine 515 appears to be linked to transcriptional activity and protein turnover [13] and serine 578 is linked to nuclear translocation and association with coactivators [14]. Phosphorylation of serine 650, within the hinge region between the DNA- and ligand-binding domains, increases nuclear export of AR and is phosphorylated in response to stress and androgen signaling [15, 16]. Several established methods have now been developed to determine the phosphorylated state of a protein both in vitro and in vivo and the production of phospho-specific antibodies has been instrumental in correlating specific modifications with protein function. Using the AR as an example we illustrate the use of a panel of receptor-specific phospho-antibodies directed against serines 81, 210/213, 308, and 650 (Fig. 1a). We will present methodologies for immunoprecipitation, western blotting, and immunohistochemistry (IHC) of the AR and phosphorylated forms of the receptor.

2 Materials

2.1 Cell and Tissue Lysis Components

- Lysis buffer: 50 mM Tris–HCl, pH 7.4; 1% v/v Triton X100; 0.25% w/v sodium deoxycholate; 150 mM NaCl; 1 mM EDTA.
- At the time of lysis the following proteinase and phosphatase inhibitors are added to the buffer solution:
  - Proteinase inhibitors (Roche, Complete®Mini, one tablet per 10 ml of lysis buffer).
  - 1 mM PMSF (100 μl per 10 ml of buffer) (*see* **Note 1**).
  - Phosphatase inhibitors cocktail I and II (Sigma-Aldrich, Dorset, UK; #P2850 and #P5726; 100 μl of each per 10 ml of buffer).
  - Sodium fluoride and β-glycerophosphate to final concentrations of 5 mM and 2 mM, respectively. Prepare 50×

stock solution, 1 ml aliquots, and store at −20 °C. Add 200 μl per 10 ml of buffer.

– Activated sodium orthovanadate to a final concentration of 2 mM. Prepare a 400 mM stock solution (*see* **Note 2**).

2.2 Immuno-precipitation Components

- Ezview®Red protein A affinity agarose gel beads (Sigma-Aldrich, Dorset, UK; #P6486).
- Ice-cold lysis buffer (with proteinase and phosphatase inhibitors: Subheading 2.1).
- Polyclonal rabbit anti-AR antibody (PG-21, Merck Millipore, Watford, UK; #06–680; Table 1 and **Note 3**).
- Rabbit IgG.
- Novex® Tris–Glycine SDS Sample Buffer (2×) (Life Technologies, Paisley, UK; #LC2676).
- NuPAGE® Sample Reducing Agent (10×) (Life Technologies, Paisley, UK; #LC2676).

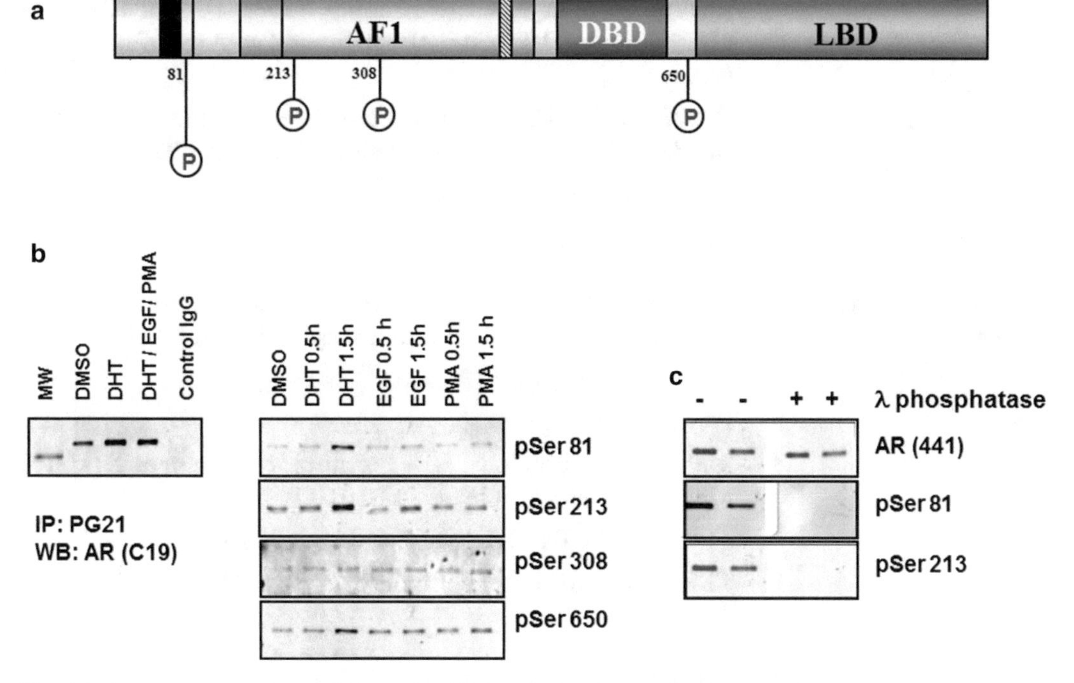

Fig. 1 Androgen receptor phosphorylation in LNCaP cells. (**a**) Schematic diagram of the androgen receptor (AR) showing the domain organization and a selection of phosphorylated serine residues: AF1, activation function 1 in the NTD; *DBD* DNA-binding domain; and *LBD* ligand-binding domain. (**b**) *Left panel*: Immunoprecipitation of the AR from LNCaP whole-cell lysate. Cells were treated with vehicle (DMSO), 5α-dihydrotestosterone (DHT, 10 nM) or DHT/epidermal growth factor (EGF, 100 ng/ml), and phorbol myristate acetate (PMA, 100 nM). *Right panel*: Detection of phosphorylated AR using phospho-specific antibodies for serines 81, 213, 308, and 650. LNCaP cells were treated with vehicle alone, DHT, EGF, or PMA for the times indicated. (**c**) The specificity of anti-phospho-serine 81 and 213 antibodies was confirmed by treating cell lysate with λ phosphatase prior to electrophoresis and western blotting

Table 1

Antibody dilutions for immunoblotting

Antibody	Working dilution	Vendor/catalogue number
Anti-AR antibody (PG-21)	1:500	Merck Millipore, Watford, UK; #06–680
Anti-phospho (Ser) 81 AR	1:500	Merck Millipore, Watford, UK; #07–1375
Anti-phospho (Ser) 213/210 AR	1:1000	Abcam, Cambridge, UK; ab71948
Anti-phospho (Ser) 308 AR	1:300	Santa Cruz Biotechnology Inc., USA; sc-26406-R
Anti-phospho (Ser) 650 AR	1:250	Abcam, Cambridge, UK; ab47563

2.3 SDS-Polyacrylamide Gel Components

- 4–12% Gradient precast Novex® 4–12% Tris–Glycine Mini Protein Gels, 1.0 mm, 12 well.
- 10× Tris–glycine buffer: 30.28 g Tris(hydroxymethyl)aminomethane (Sigma-Aldrich, Dorset), 144.3 g glycine (Sigma-Aldrich), and deionized water total volume 1 l. Do not adjust pH and store at 4 °C. 1× running buffer: 100 ml 10× Tris–glycine buffer, 10 ml of 10% SDS, and 890 ml of deionized water.
- Protein molecular weight markers: SeeBlue® Plus2 prestained standard (Life Technologies, Paisley, UK; #LC5925).
- Gel electrophoresis apparatus.
- Constant voltage power supply.

2.4 Immunoblotting Components

- 1× Transfer buffer: 100 ml of 10× Tris–glycine buffer, 200 ml of methanol, and 700 ml water, keep at 4 °C until use.
- Whatman filter paper (3 mm).
- Western blot apparatus.
- Millipore Immobilon-FL western blot membranes (Merck Millipore, Watford, UK; #IPFL10100).
- TBS: 50 mM Tris–HCl pH 7.5, 150 mM NaCl.
- TBST: 50 mM Tris–HCl pH 7.5, 150 mM NaCl, 0.1% v/v Tween 20.

2.5 Immunohistochemistry Components

- Xylene.
- Absolute alcohol (100% ethanol).
- 95% v/v Ethanol solution.
- 90% v/v Ethanol solution.
- 70% v/v Ethanol solution.
- 0.01 M Citrate buffer pH = 6.0.

- 3% v/v Hydrogen peroxide in methanol solution.
- Normal goat serum.
- TBS: 50 mM Tris–HCl pH 7.5, 150 mM NaCl.
- Bovine serum albumin (BSA).
- EnVision+ System-HRP (DAB+) system (DAKO UK, Ltd., Cambridge, UK; #K4010 and #K4006).
- Hematoxylin solution, Harris Modified (Sigma-Aldrich, Dorset, UK).
- 1% v/v Acid alcohol solution (1 ml of HCl mixed with 70% v/v ethanol solution up to 100 ml).
- Scott's tap water (0.2% ammonia water: 2 ml of ammonium hydroxide in final volume of 1 l of water).

2.6 Immunofluorescence Components

- Xylene.
- Absolute alcohol (100% ethanol).
- 95% v/v Ethanol solution.
- 90% v/v Ethanol solution.
- 70% v/v Ethanol solution.
- 0.01 M Citrate buffer pH = 6.0 prepare form 0.1 M stock.
- 3% v/v Hydrogen peroxide in methanol solution.
- Normal goat serum.
- TBS: 50 mM Tris–HCl pH 7.5, 150 mM NaCl.
- Bovine serum albumin (BSA).
- EnVision+ System-HRP (DAB+) system (DAKO UK, Ltd., Cambridge, UK; #K4010 and #K4006).
- TSA™ Plus Fluorescence Systems (Perkin Elmer Inc, Waltham, US; #NEL 754).
- ProLong® Gold Antifade Reagent with DAPI (Life Technologies, Paisley, UK; # P-36931).

2.7 Antibodies (Tables 1–3 and Note 3)

- Polyclonal rabbit anti-AR antibody (PG-21, Merck Millipore, Watford, UK; #06-680).
- Polyclonal rabbit anti-AR antibody (C-19, Santa Cruz Biotechnology Inc., US; sc-815).
- Polyclonal rabbit anti-AR antibody (N-20, Santa Cruz Biotechnology Inc., US; sc-816).
- Monoclonal mouse anti-AR antibody (AR411, Abcam, Cambridge, UK; #ab9474).
- Anti-phospho 81 AR (Merck Millipore, Watford, UK, #07-1375).
- Anti-phospho 213/210 AR (Abcam, Cambridge, UK; # ab71948).

- Anti-phospho Ser 308 (Santa Cruz Biotechnology Inc., US; #sc-26406-R).
- Anti-phospho 650 AR (Abcam, Cambridge, UK; # ab47563).
- Secondary antibodies: Goat anti-Rabbit IgG (H + L) IRDye 800CW, LI-COR Biosciences Ltd., USA; #926-32211 and Goat anti-Mouse IgG (H + L) IRDye® 680RD, LI-COR Biosciences Ltd., USA #926-68070).
- EnVision+ System-HRP (DAB+) system (DAKO UK, Ltd., Cambridge, UK; #K4010 and #K4006).

3 Methods

Phosphorylation of nuclear receptors plays an important role in regulating receptor stability, intracellular localization and interactions with co-regulatory proteins and DNA. Using phospho-specific antibodies it is now possible to correlate specific phosphorylated residues with receptor function and to screen cell and tissue samples for tissue-specific phosphorylation or for changes in disease. In the following sections we describe methods for studying the phosphorylation of the AR in a cell culture model, in response to different hormone or growth factor treatments. We then describe using phospho-specific antibodies to visualize the receptor in tissue samples by IHC and immunofluorescence microscopy.

3.1 Cell lysis

- Wash cells twice with PBS without Ca^{2+} or Mg^{2+}, followed by two washes with ice-cold PBS without Ca^{2+} or Mg^{2+} supplemented with phosphatase inhibitors (5 mM sodium fluoride, 2 mM β-glycerophosphate, and 2 mM activated sodium orthovanadate).
- Lyse the cells on the cell culture plate by adding 50 μl of ice-cold lysis buffer per 1×10^6 cells. Disrupt and disintegrate the cells using a cell scraper (*see* **Note 4**), collect, and spin at 10,000 rpm for 4 min. Collect a 15 μl aliquot for protein-level determination and store the rest of the lysate at −80 °C until use.

3.2 Immunoprecipitation

- Aliquot 50 μl of Ezview® Red protein A affinity agarose beads into a clean tube on ice (*see* **Note 5**). Wash the protein A agarose twice with 750 μl of ice-cold lysis buffer, vortex, and spin at 10,000 rpm for 30 s at 4 °C in a benchtop centrifuge.
- Pre-absorb cell lysate by adding cell lysate (500 μl) to washed beads, vortex briefly, and incubate at 4 °C with gentle agitation for 40–60 min. Centrifuge for 30 s at 10,000 rpm in a benchtop centrifuge, remove supernatant and keep on ice, and discard pre-absorption beads.
- Add 5 μl of polyclonal rabbit anti-AR antibody (PG-21) for every 0.5 mg/ml of pre-absorbed cell lysate suspension and

increase volume to 0.5 ml (if necessary) using ice-cold PBS. Vortex briefly and incubate at 4 °C with gentle agitation overnight; centrifuge for 30 s at 10,000 rpm in a benchtop centrifuge. Add this mixture to 50 μl of freshly washed Ezview® Red protein A agarose beads; vortex briefly and incubate at 4 °C with gentle agitation for 1–2 h. Centrifuge for 30 s at 10,000 rpm in a benchtop centrifuge. Then wash bead pellet three times using 750 μl of ice-cold lysis buffer, vortex briefly, incubate on ice with gentle agitation for 5 min, and then centrifuge for 30 s at 10,000 rpm in a benchtop centrifuge.

- Elute antibody-antigen complex from the agarose beads by adding 50 μl of Novex® Tris–glycine SDS sample buffer (2×) containing 50 mM dithiothreitol (NuPAGE® sample reducing agent (10×)); vortex briefly and incubate for 5–10 min at 95 °C. Centrifuge for 30 s at 10,000 rpm in a benchtop centrifuge (*see* **Note 6**). Remove supernatant and use immediately or store at −80 °C until use (*see* **Note 7**).

3.3 4–12% Gradient Sodium Dodecyl Sulfate-Polyacrylamide Gel Electrophoresis

- Protocol based on the system developed by Laemmli.
- Precast gradient gels are mounted in the electrophoresis apparatus and 1× running buffer was added to cover the gels completely. 5 μl of markers were added to one well. Samples were then loaded into subsequent wells (10–20 μl per well depending on protein concentration in the samples).
- The gel apparatus cover was then attached and a continuous voltage (150 V) was passed through the gel for approximately 1.5 h, or until the dye front reached the bottom of the gel (*see* **Note 8**). Once electrophoresis has been completed the gel plates are prised open and the gel is rinsed with water and placed gently in a container with 1× transfer buffer.

3.4 Immunoblotting

Pre-soak membrane in methanol (1 min) prior to use and then equilibrate in transfer buffer for at least 5–10 min together with Whatman filter papers (*see* **Note 9**). Assemble the gel/membrane sandwich (sponge, 3× filter paper, gel, membrane, 3× filter paper, sponge) and place sandwich into transfer tank and transfer at a constant voltage of 65 V for 3 h (*see* **Notes 10** and **11**).

- After transfer remove membrane from sandwich (*see* **Note 12**), carefully using tweezers, place in a container with TBS for 10–15 min with gentle agitation, and then place membrane in freshly prepared blocking solution; this is 5% w/v nonfat skimmed milk in TBST for most antibodies; however the 5% w/v nonfat skimmed milk is replaced with 5% w/v casein for phospho-specific antibodies; incubate for 1 h at room temperature. Remove blocking solution and add the primary antibody (anti-AR or anti-phospho AR, *see* Table 1 for a list of the antibodies used and dilutions) diluted in the fresh blocking solution appropriate for the antibody. Incubate overnight at 4 °C with gentle agitation.

- Wash membranes 3–4 times with TBST, 5 min per wash (*see* **Note 12**).
- Add secondary antibody (1:10,000) in TBST containing SDS (1 μl of 5 % SDS per 30 ml of antibody diluent), and incubate for 1 h at room temperature. Wash membrane twice in TBST, followed by twice in TBS (*see* **Note 12**). Visualize membrane with an imaging system; the example presented here used the Li-COR Odyssey (Fig. 1b).

3.5 Immunohistochemistry

This protocol can be applied to PFA-fixed tissue samples (*see* **Note 13**) or commercially available tissue micro arrays (e.g. the array used to generate these methods was AccuMax® arrays: prostate cancer tissue stages II and III (A222, A223)).

- Dewax and rehydrate paraffin-embedded sections by washing sequentially in xylene (twice for 7 min), absolute alcohol (twice for 2 min), 95 % v/v ethanol solution (twice for 2 min), 90 % v/v ethanol solution (twice for 2 min), and 70 % v/v ethanol solution (twice for 2 min) and rinse in water for 5–10 min.
- Antigen retrieval: Prepare 0.1 M citrate buffer solution pH = 6.0, place it in an open pressure cooker, bring to boil, carefully place rack with slides, close cooker, and cook slides under full pressure for 5 min. Remove pressure cooker from the heat and release pressure. Leave it to cool in citrate buffer until it reaches room temperature. Rinse with water for 5–10 min.
- Remove endogenous peroxidase activity using a 3 % v/v hydrogen peroxide in methanol solution for 30 min at room temperature. Rinse in water for 5–10 min followed by wash in TBST. Block nonspecific antibody binding with 20 % v/v normal goat serum in TBST with 5 % bovine serum albumin (BSA) (NGS/TBST/BSA) for 1 h at room temperature (*see* **Note 14**). Drain liquid from the slide and carefully remove excess of fluid.
- Add primary antibody diluted in fresh NGS/TBST/BSA solution, incubate overnight at 4 °C (*see* Table 2 for a list of antibodies and dilutions), and remember to include negative control slide in each run. All antibody specificities were validated using blocking peptides [17] and λ protein phosphatase (100 U at 37 °C) (*see* Fig. 1c). Remove primary antibody solution from slides and wash them twice (5 min) with TBST.
- Add secondary antibody HRP conjugated (EnVision+ System-HRP (DAB+) and incubate for 30 min at room temperature. Wash slides twice (5 min) in TBST followed by single 1–2-min wash in TBS. To develop chromogenic reactions add DAB substrate (one drop per 1 ml of chromogenic buffer solution) to

Table 2
Antibody dilutions for IHC

Antibody	Working dilution	Vendor/catalogue number
Anti-AR antibody (N-20)	1:250	Santa Cruz Biotechnology Inc., USA; sc-816
Anti-phospho (Ser) 81 AR	1:50	Merck Millipore, Watford, UK; #07–1375
Anti-phospho (Ser) 213/210 AR	1:25	Abcam, Cambridge, UK; ab71948
Anti-phospho (Ser) 308 AR	1:200	Santa Cruz Biotechnology Inc., USA; sc-26406-R
Anti-phospho (Ser) 650 AR	1:25	Abcam, Cambridge, UK; ab47563

Table 3
Antibody dilutions for immunofluorescence (*see* Note 13)

Antibody	Working dilution	Vendor/catalogue number
Anti-phospho (Ser) 81 AR	1:500	Merck Millipore, Watford, UK; #07–1375
Anti-phospho (Ser) 213/210 AR	1:250	Abcam, Cambridge, UK; ab71948
Anti-phospho (Ser) 650 AR	1:250	Abcam, Cambridge, UK; ab47563

each and monitor color development using microscope. Halt color development by washing with water, and then counterstain using Harris hematoxylin (hematoxylin—45 s/wash in water/dip 1–2 times in acidic alcohol solution, wash in water, then in Scott's tap water for 25–30 s, wash in water). Dehydrate slides by immersing in serial alcohol solutions 70, 80, and 95% for 5–10 s each, followed by two washes in absolute alcohol for about 20 s each. Finally wash twice for 5 min each with xylene. Mount the slides with Pertex: example staining of normal and tumorigenic prostate tissue is shown in Fig. 2.

3.6 Immuno-fluorescence

- For slides and antigen retrieval follow the same protocol as IHC, as well as peroxidase reduction.
- Block nonspecific antibody binding with 20% v/v normal goat serum in TBST with 5% w/v bovine serum albumin (BSA) (NGS/TBST/BSA) for 30 min at room temperature (*see* **Note 14**). Drain liquid from the slide and carefully remove excess of fluid.
- Add first primary antibody diluted in fresh NGS/TBST/BSA solution anti-AR antibody (N20, 1:1000 dilution; Santa Cruz Biotechnology Inc., USA; #sc-816), and incubate overnight at 4 °C. Remove primary antibody solution from slides and wash them three times for 5 min each with TBST.

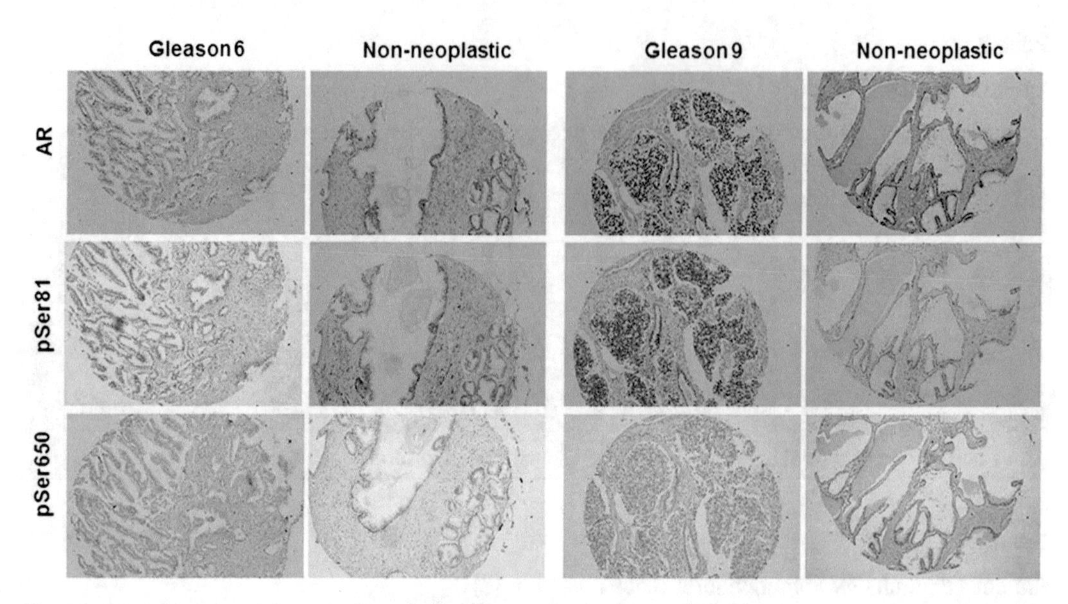

Fig. 2 Immunohistochemistry detection of the AR in prostate tissue samples. A prostate tissue microarray was stained for either total AR or receptor phosphorylated on serine 81 or serine 650. The figure shows examples of neoplastic prostate cancer (Gleason Score 6 and 9) and matched non-neoplastic tissue. Increased staining of total and phospho-serine 81 in the cancer samples was observed

- Add ready-to-use HRP-conjugated secondary antibody. Remove secondary antibody solution from slides and wash them three times for 5 min each with TBST.
- Add tyramide signal amplification reagent (TSA™ Plus Fluorescence Systems; Perkin Elmer Inc, Waltham, USA; #NEL 754) working solution (Cy3-red) (1:50 dilution) for 10 min at room temperature. Wash three times for 5 min each in TBST. Check staining under microscope.
- Perform second antigen retrieval (bring to boil citrate buffer in microwave and place slides for 2.5 min at max power, make sure that slides are covered with solution). Bring them back to room temperature, and wash with deionized water followed by two washes in TBST.
- Block again for 30 min at room temperature and incubate with second primary antibody directed against phosphorylated AR (Table 3 and **Note 15**) diluted in fresh blocking solution overnight at 4 °C. Remove the antibody solution from slides and wash them three times for 5 min each with TBST.
- Add ready-to-use HRP-conjugated secondary antibody. Remove secondary antibody solution from slides and wash them three times for 5 min each with TBST.
- Add tyramide signal amplification reagent (TSA™ Plus Fluorescence Systems; Perkin Elmer Inc, Waltham, USA; #NEL 754) working solution (fluorescein-green) (1:50 dilu-

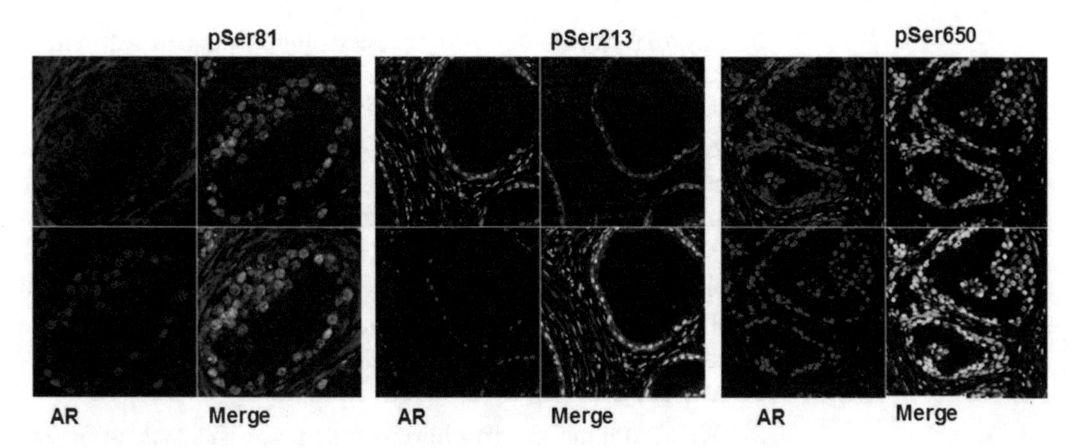

Fig. 3 Double-immunofluorescence staining showing colocalization of AR and phosphorylated AR in prostate tissue. Transurethra resection of prostate (TURP) sample sections was stained for total AR (red) or specific phosphorylated serines (81, 213, and 650) (*green*). Nuclear receptor staining can be seen in the glandular epithelial cells

tion) for 10 min at room temperature. Wash three times for 5 min in TBST.

- Counterstain slides with DAPI to visualize nuclei and mount slides (ProLong® Gold Antifade Reagent with DAPI; Life Technologies, Paisley, UK; # P-36931); colocalization of total AR and phospho-AR in human prostate tissue is illustrated in Fig. 3.

4 Notes

1. The half-life for PMSF is 55 min at pH 7.5, 4 °C; this should be borne in mind when preparing the buffer in advance; that is, PMSF is best added just before lysis to maximize performance.
2. Preparation of activated sodium orthovanadate solution:

   Prepare a solution of 800 mM sodium orthovanadate in water and adjust pH to 10 with concentrated NaOH. The color of the solution will change to yellow. Boil the solution, till it becomes colorless, and readjust pH to 10. Repeat boiling and pH adjustment steps until solution remains colorless. Add water to obtain a 400 mM stock. Make 1 ml aliquots of solution and store at −20 °C until needed. If sodium orthovanadate solution appears cloudy warm it up in a water bath until the solution becomes clear and then use.
3. Antibodies can vary from batch to batch; therefore some optimization may be required every time a new batch is acquired.
4. For cell lysis scrape hard, ~20–30 moves in different directions; there is no need to be gentle as you want to release cell contents.

5. Ensure that Ezview® slurry is uniformly suspended before use. Remove the end of the pipette tip (approximately 2 mm) to allow the agarose bead suspension to be accurately pipetted.
6. Retain samples of supernatants and washes from the immunoprecipitation to monitor antigen retrieval/loss. Controls for immunoprecipitation can be run with specific blocking peptides or an irrelevant antibody.
7. Purified AR must be used or stored at –80 °C immediately after preparation as it can be very unstable [17]. DHT can be added to the lysis buffer to stabilize the molecule (final concentration 10^{-8} M).
8. When running electrophoresis gels place the tank in an ice bath to reduce running temperature.
9. Nitrocellulose membranes can also be used; however do not pre-soak in methanol as it can damage the membrane.
10. Ensure that all air bubbles are removed from the gel/membrane sandwich by gently rolling with a pencil or round tube, ensuring that you do not disturb the gel.
11. Remember that proteins in the samples now have a negative charge due to the SDS usage, so be careful when you are starting transfer. If electrodes are attached with the wrong polarity your proteins will end up in the transfer buffer instead of on the membrane.
12. Never let the membrane dry as this will cause problems with the analysis.
13. Local and national ethical permissions and rules must be adhered to when using patient samples.
14. For IHC, secondary blocking normal goat serum is used because secondary antibodies are from goat; if the secondary antibodies used are from a different species the serum utilized at this stage should be altered to match.
15. Be aware that labeled antibodies are light sensitive. Therefore keep them covered when incubating by covering your container with foil or by using a covered western incubation box.

Acknowledgements

This work was carried out at the former MRC Human Reproductive Sciences Unit, Edinburgh. The work was supported by a visiting scientist grant (I.J.M.) and postdoctoral fellowship (D.McG.) from the Medical Research Council, UK. We are grateful to Dr. Axel Thomson for providing tissue samples.

References

1. Rubin CS, Rosen OM (1975) Protein phosphorylation. Annu Rev Biochem 44:831–887
2. van der Steen T, Tindall DJ, Huang H (2013) Posttranslational modification of the androgen receptor in prostate cancer. Int J Mol Sci 14(7):14833–14859
3. Coffey K, Robson CN (2012) Regulation of the androgen receptor by post-translational modifications. J Endocrinol 215(2):221–237
4. Anbalagan M, Huderson B, Murphy L, Rowan BG (2012) Post-translational modifications of nuclear receptors and human disease. Nucl Recept Signal 10:e001
5. Chen S, Xu Y, Yuan X, Bubley GJ, Balk SP (2006) Androgen receptor phosphorylation and stabilization in prostate cancer by cyclin-dependent kinase 1. Proc Natl Acad Sci U S A 103(43):15969–15974
6. Gordon V, Bhadel S, Wunderlich W, Zhang J, Ficarro SB, Mollah SA, Shabanowitz J, Hunt DF, Xenarios I, Hahn WC, Conaway M, Carey MF, Gioeli D (2010) CDK9 regulates AR promoter selectivity and cell growth through serine 81 phosphorylation. Mol Endocrinol 24(12):2267–2280
7. Chen S, Gulla S, Cai C, Balk SP (2012) Androgen receptor serine 81 phosphorylation mediates chromatin binding and transcriptional activation. J Biol Chem 287(11):8571–8583
8. Ward RD, Weigel NL (2009) Steroid receptor phosphorylation: assigning function to site-specific phosphorylation. Biofactors 35(6):528–536
9. Taneja SS, Ha S, Swenson NK, Huang HY, Lee P, Melamed J, Shapiro E, Garabedian MJ, Logan SK (2005) Cell-specific regulation of androgen receptor phosphorylation in vivo. J Biol Chem 280(49):40916–40924
10. Lin HK, Yeh S, Kang HY, Chang C (2001) Akt suppresses androgen-induced apoptosis by phosphorylating and inhibiting androgen receptor. Proc Natl Acad Sci U S A 98(13):7200–7205
11. Shu SK, Liu Q, Coppola D, Cheng JQ (2010) Phosphorylation and activation of androgen receptor by Aurora-A. J Biol Chem 285(43):33045–33053
12. Ponguta LA, Gregory CW, French FS, Wilson EM (2008) Site-specific androgen receptor serine phosphorylation linked to epidermal growth factor-dependent growth of castration-recurrent prostate cancer. J Biol Chem 283(30):20989–21001
13. Chymkowitch P, Le May N, Charneau P, Compe E, Egly JM (2010) The phosphorylation of the androgen receptor by TFIIH directs the ubiquitin/proteasome process. EMBO J 30(3):468–479
14. Koryakina Y, Ta HQ, Gioeli D (2014) Androgen receptor phosphorylation: biological context and functional consequences. Endocr Relat Cancer 21(4):T131–T145
15. Clinckemalie L, Vanderschueren D, Boonen S, Claessens F (2012) The hinge region in androgen receptor control. Mol Cell Endocrinol 358(1):1–8
16. Gioeli D, Black BE, Gordon V, Spencer A, Kesler CT, Eblen ST, Paschal BM, Weber MJ (2006) Stress kinase signaling regulates androgen receptor phosphorylation, transcription, and localization. Mol Endocrinol 20(3):503–515
17. McEwan IJ, McGuinness D, Hay CW, Millar RP, Saunders PT, Fraser HM (2010) Identification of androgen receptor phosphorylation in the primate ovary in vivo. Reproduction 140(1):93–104

Chapter 8

Mapping Protein–DNA Interactions Using ChIP-exo and Illumina-Based Sequencing

Stefan J. Barfeld and Ian G. Mills

Abstract

Chromatin immunoprecipitation (ChIP) provides a means of enriching DNA associated with transcription factors, histone modifications, and indeed any other proteins for which suitably characterized antibodies are available. Over the years, sequence detection has progressed from quantitative real-time PCR and Southern blotting to microarrays (ChIP-chip) and now high-throughput sequencing (ChIP-seq). This progression has vastly increased the sequence coverage and data volumes generated. This in turn has enabled informaticians to predict the identity of multi-protein complexes on DNA based on the overrepresentation of sequence motifs in DNA enriched by ChIP with a single antibody against a single protein. In the course of the development of high-throughput sequencing, little has changed in the ChIP methodology until recently. In the last three years, a number of modifications have been made to the ChIP protocol with the goal of enhancing the sensitivity of the method and further reducing the levels of nonspecific background sequences in ChIPped samples. In this chapter, we provide a brief commentary on these methodological changes and describe a detailed ChIP-exo method able to generate narrower peaks and greater peak coverage from ChIPped material.

Key words Chromatin immunoprecipitation, Exonuclease, Cancer, Androgen receptor, Prostate

1 Introduction

In 1988, Solomon et al. first described the mapping of protein–DNA interactions in vivo using formaldehyde as a cross-linking agent to stabilize these otherwise fragile interactions [1]. Ever since then, the chromatin immunoprecipitation (ChIP) protocol has only moderately changed, with the introduction of ultrasonic shearing devices and magnetic beads for separation being two notable exceptions. However, the downstream analysis of precipitated DNA fragments has been subject to immense changes and developments. Initially, the introduction of the ChIP-chip microarray technique allowed a first global and unbiased overview of the precipitated material [2]. Although this approach was a groundbreaking development, it was still far from optimal as it suffered

Iain J. McEwan (ed.), *The Nuclear Receptor Superfamily: Methods and Protocols*, Methods in Molecular Biology, vol. 1443, DOI 10.1007/978-1-4939-3724-0_8, © Springer Science+Business Media New York 2016

from high noise, delivered relatively low resolution, and was prone to artifacts. With the emergence of affordable high-throughput sequencing, first global ChIP-seq studies for both histone marks and transcription factors were published in 2007 [3, 4]. The main advantage of ChIP-seq over the ChIP-chip technology is undoubtedly the resolution but other factors, such as the possibility to multiplex (i.e., simultaneous sequencing of multiple samples in a single lane using unique indices/barcodes) and reduced bias, should not be discounted.

However, the ability to quickly generate tens, and up to hundreds, of million reads per sample also imposes certain challenges on researchers, as these data require both storage space and novel computational approaches to interpret. As of today, there is no obligatory analysis pipeline or quality control when publishing ChIP-seq studies. Thus, studies vary in their sequencing depth and quality as well as algorithms used for alignment and peak calling. In an effort to define uniform guidelines and make ChIP-seq experiments more reproducible between different labs, the ENCODE consortium recently published an updated version of their recommendations, which should be regarded as the gold standard [5].

In recent years, several minor and major modifications of the original ChIP-seq approach have emerged. These approaches generally try to tackle the biggest issue of conventional ChIP-seq, the requirement for large number of cells. We have compiled a list of the most popular modifications and listed their advantages over regular ChIP-seq (Table 1) [6–12].

Among these, the incorporation of 5′–3′ strand-specific exonuclease digestion into the standard ChIP-seq procedure, ChIP-exonuclease (ChIP-exo), stands out as it does not necessarily aim at reducing input material but rather at improving signal-to-noise ratio, increasing plexity and cutting down hands-on time. ChIP-exo refines the conventional ChIP-seq by performing several on-bead enzymatic reactions prior to elution of the ChIPped DNA, thereby eliminating most of the time-consuming AMPure XP bead cleanups (Fig. 1). In addition, two on-bead exonuclease reactions digest DNA fragments until they reach protected areas (Fig. 2). This greatly reduces fragment length and noise and allows for more multiplexing as it lowers the requirements for sequencing reads necessary for robust peak calling. In addition, ChIP-exo facilitates the high resolution of TF-binding locations and binding motifs by significantly reducing peak widths (Fig. 2) [11]. Motif enrichment analysis of ChIP peaks remains a key tool in inferring the composition of multi-protein transcription complexes.

The method was originally developed by Rhee et al. in 2011 for the analysis of yeast transcription factor-binding sites on the SOLID sequencing platform (Applied Biosystems) [11]. Recently, however, Serandour et al. developed an Illumina-based counterpart [12],

Table 1
Recent modifications to ChIP-seq methodology

Method	Adaptation	Sensitivity	Advantages/disadvantages	Target proteins	References
Linear amplification of DNA (LinDA)	In vitro transcription instead of PCR, single-tube reactions	5,000–10,000 cells	Reduced PCR induced bias, higher sensitivity	TF (ERα) and histones (H3K4me3)	[6]
Micro/nano ChIP-seq	Various, e.g., single-tube reactions, restriction enzymes	~10,000 cells	Higher sensitivity	TF (CTCF) and histones (various)	[7, 8]
PAT-ChIP (pathology tissue)	ChIP-seq of formalin-fixed, paraffin-embedded (FFPE) samples, includes deparaffinization and MNase digestion	10 μM FFPE sections	Higher sensitivity, allows ChIP-seq from FFPE samples	TF (CTCF), Pol2 and histones (various)	[9]
Carrier-assisted ChIP	Histones or mRNA as carrier to facilitate immunoprecipitation and degraded prior to library preparation	10,000	Used on core needle biopsy samples, high sensitivity	TF (ERα)	[10]
ChIP-exo	Additional exonuclease digestion of unprotected DNA straightens peaks and reduces noise	Not tested, several millions used	Higher plexity, narrower peaks, reduced noise	TF (ERα), not applicable for histones	[11, 12]

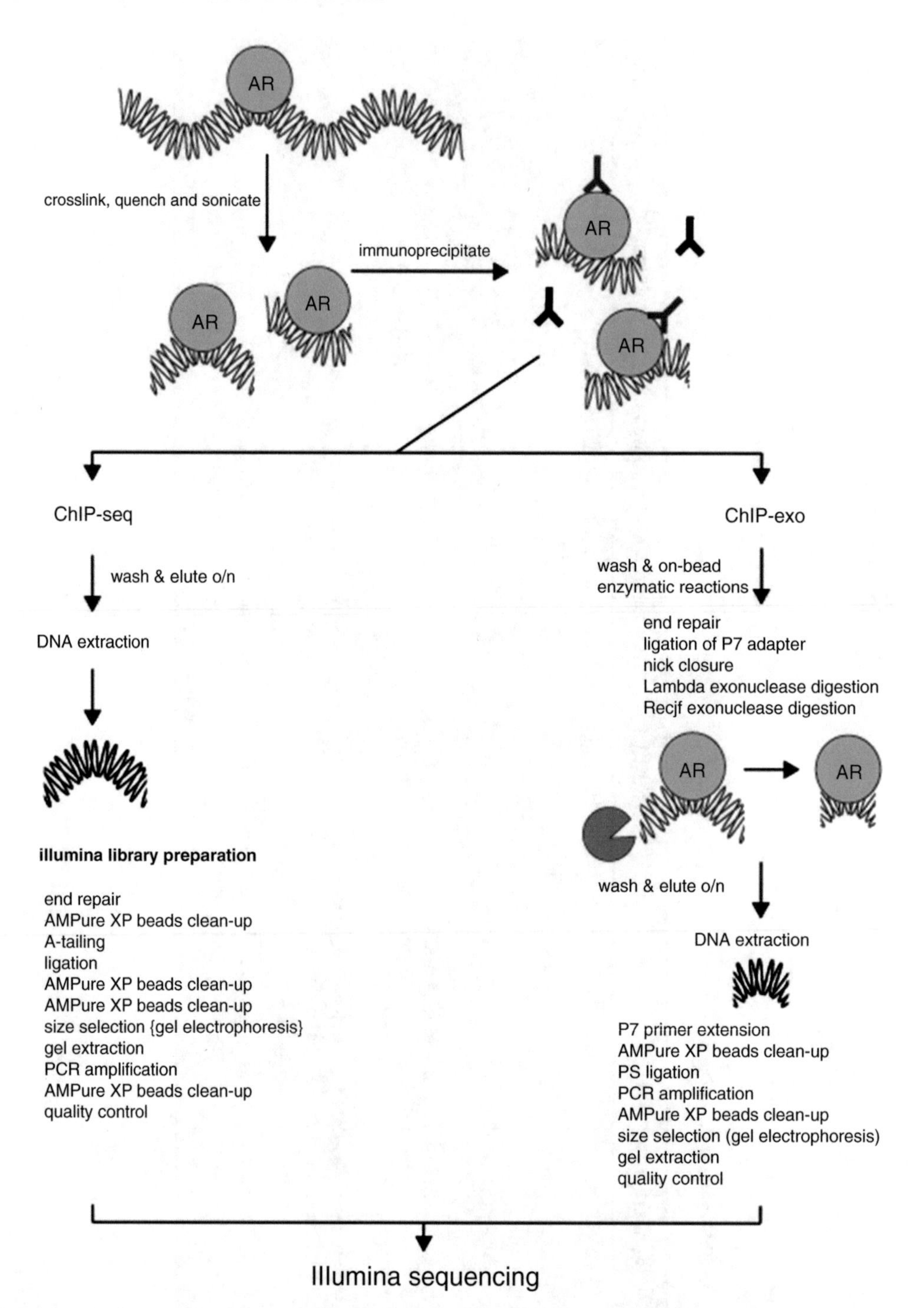

Fig. 1 Schematic workflow comparison of ChIP-seq and ChIP-exo from immunoprecipitation to DNA extraction. ChIP-exo and ChIP-seq share the initial steps including cross-linking, quenching, sonication, and immunoprecipitation. Subsequently, in the conventional ChIP-seq protocol the precipitated DNA is eluted and extracted prior to Illumina TruSeq library preparation. In the ChIP-exo protocol, however, the precipitated DNA remains linked to the magnetic beads. This allows rapid washing and on-bead enzymatic reactions, including two novel exonuclease digestions that reduce fragment length and background, which eliminates many of the time-consuming AMPure XP bead cleanups

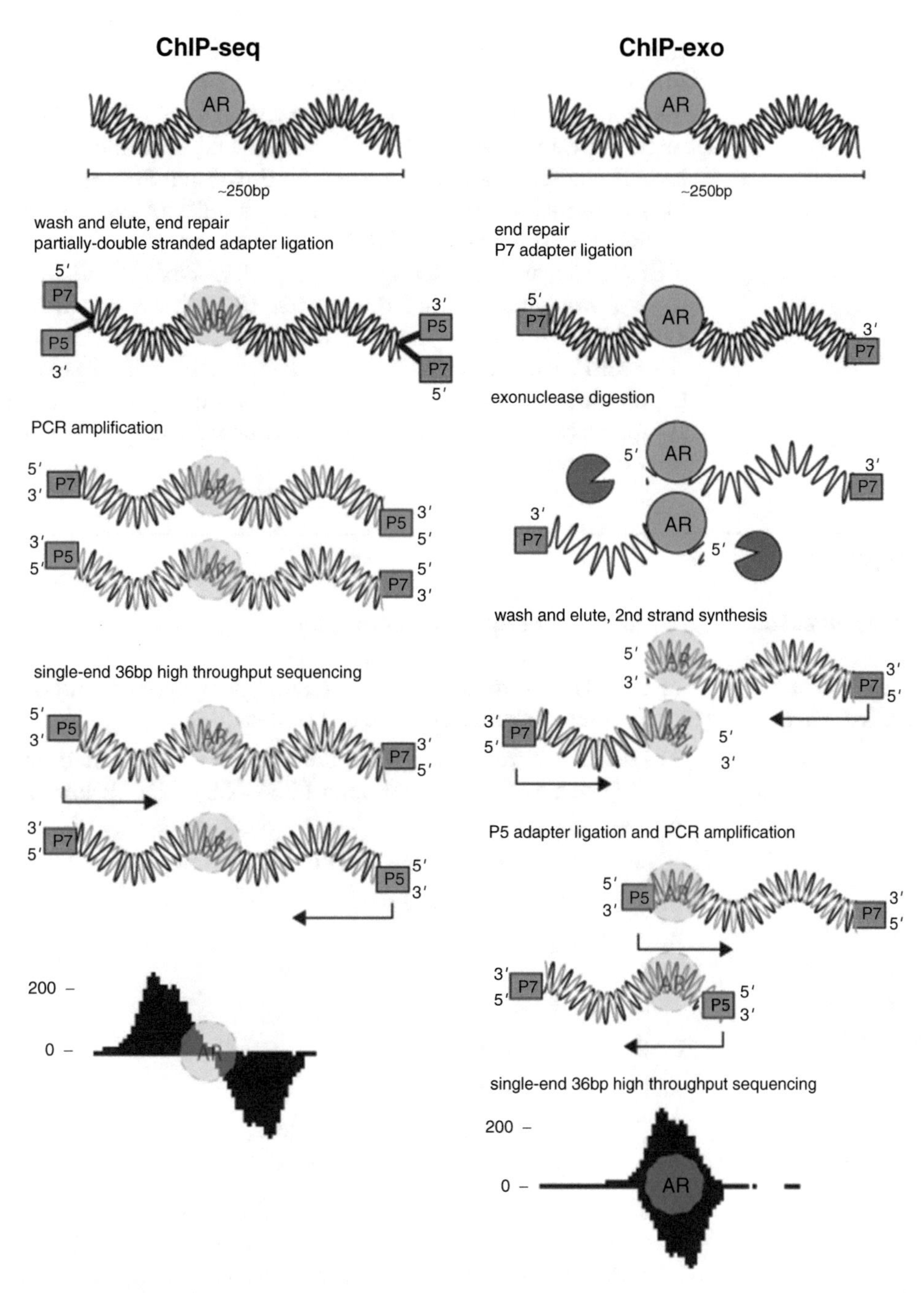

Fig. 2 Preparation of ChIPped DNA for sequencing using a conventional protocol or ChIP-exo. In the traditional Illumina-based library preparation, partially double-stranded adapters are ligated to the extracted DNA. PCR amplification and subsequent single-ended high-throughput sequencing result in two shifted populations of reads, one derived from the original top strand, and the other one from the original bottom strand. In the ChIP-exo protocol, however, the novel exonuclease reaction digests DNA until it reaches protected DNA (i.e., shielded transcription factor-binding sites). This results in two overlapping populations of sequencing reads as the sequencing adapters are ligated directly up- and downstream of the protected areas. Thus, ChIP-exo allows increased multiplexing through better coverage of the actual transcription factor-binding sites and generates narrower peak widths, which improves motif enrichment analysis

thereby enabling it to be compatible with the currently most widely used sequencing platform. Here, we focus on this application of this method to study androgen receptor (AR) binding in prostate cancer cells since ChIP-seq has become a vital instrument to define the AR transcriptome and several landmark studies have utilized this technique. These include ChIP-seq studies of the AR itself in cell lines and tumor tissue [13, 14], ETS transcription factors, such as ERG [15], and pioneering factors, such as FoxA1 [16].

We provide a focused description of positive- and negative-control experiments to measure androgen-stimulated AR binding using ChIP-exo in combination with direct Illumina-based high-throughput sequencing. However, this method is also more generally applicable to the study of AR and other transcription factors in other contexts.

2 Materials

2.1 Oligonucleotides

See Table 2 for sequence information.

2.2 Cell Culture and Cross-Linking

1. RPMI1640 media (Life Technologies) supplemented with 10% fetal bovine serum (FBS) (Gibco).
2. Phenol red-free RPMI (Life Technologies), supplemented with 10% charcoal dextran-stripped FBS (CSS) (Life Technologies).
3. AR ligands, for example 5α-dihydrotestosterone/5α-androstan-17β-ol-3-one (DHT) (Sigma) or R1881 (Sigma).
4. Formaldehyde (e.g., Sigma).
5. 2.5 M Glycine solution.

2.3 Harvesting, Lysis, and Sonication of Cells

1. Cell scrapers.
2. PBS (Life Technologies).
3. Rotating tube mixer at 4 °C.
4. Diagenode Bioruptor (Diagenode).
5. Proteinase inhibitor cocktail (Roche).
6. Lysis buffer 1 (LB1): 50 mM HEPES–KOH (pH 7.5), 140 mM NaCl, 1 mM EDTA, 10% glycerol, 0.5% NP-40, 0.25% Triton-X—add 1× proteinase inhibitor cocktail fresh before use.
7. Lysis buffer 2 (LB2): 10 mM Tris–HCl (pH 7.5), 200 mM NaCl, 1 mM EDTA, 0.5 mM EGTA—add 1× proteinase inhibitor cocktail fresh before use.
8. Lysis buffer 3 (LB3): 10 mM Tris–HCl (pH 7.5), 100 mM NaCl, 1 mM EDTA, 0.5 mM EGTA, 0.1% sodium-deoxycholate, 0.5% SDS—add 1× proteinase inhibitor cocktail fresh before use.
9. 10% Triton-X in LB3.

Table 2
Oligonucleotide sequences

Primer	Sequence
P7 exo-adapter reverse (5′ Phos = phosphorylated 5′ end)	5′ Phos-GTGACTGGAGTTCAGACGTGTGCTCTTCCGATC-OH 3′
P7 exo-adapter forward	5′ OH-GATCGGAAGAGCACACGTCT-OH 3′
P5 exo-adapter reverse	5′ OH-AGATCGGAAGAGCG-OH 3′
P5 exo-adapter forward	5′ OH-TACACTCTTTCCCTACACGACGCTCTTCCGATCT-OH 3′
P7 primer	5′ OH-GACTGGAGTTCAGACGTGTGCT-OH 3′
PCR primer universal reverse (* = Phosphorothioates S-linkage)	5′ OH-AATGATACGGCGACCACCGAGATCTACACTCTTTCCCTACACG*A-OH 3′
PCR primer index 2 forward (* = Phosphorothioates S-linkage)	5′ OH-CAAGCAGAAGACGGCATACGAGATACATCGGTGACTGGAGTTCAGACGTGTGC*T-OH 3′
PCR primer index 4 forward (* = Phosphorothioates S-linkage)	5′ OH-CAAGCAGAAGACGGCATACGAGATTGGTCAGTGACTGGAGTTCAGACGTGTGC*T-OH 3′
PCR primer index 5 forward (* = Phosphorothioates S-linkage)	5′ OH-CAAGCAGAAGACGGCATACGAGATCACTGTGTGACTGGAGTTCAGACGTGTGC*T-OH 3′
PCR primer index 6 forward (* = Phosphorothioates S-linkage)	5′ OH-CAAGCAGAAGACGGCATACGAGATATTGGCGTGACTGGAGTTCAGACGTGTGC*T-OH 3′
PCR primer index 7 forward (* = Phosphorothioates S-linkage)	5′ OH-CAAGCAGAAGACGGCATACGAGATGATCTGGTGACTGGAGTTCAGACGTGTGC*T-OH 3′
PCR primer index 12 forward (* = Phosphorothioates S-linkage)	5′ OH-CAAGCAGAAGACGGCATACGAGATTACAAGGTGACTGGAGTTCAGACGTGTGC*T-OH 3′
PCR primer index 13 forward (* = Phosphorothioates S-linkage)	5′ OH-CAAGCAGAAGACGGCATACGAGATTTGACTGTGACTGGAGTTCAGACGTGTGC*T-OH 3′

(continued)

Table 2
(continued)

Primer	Sequence
PCR primer index 14 forward (* = Phosphorothioates S-linkage)	5′ OH-CAAGCAGAAGACGGCATACGAGATGGAACTGTGACTGGAGTTCAGACGTGTGC*T-OH 3′
PCR primer index 15 forward (* = Phosphorothioates S-linkage)	5′ OH-CAAGCAGAAGACGGCATACGAGATTGACATGTGACTGGAGTTCAGACGTGTGC*T-OH 3′
PCR primer index 16 forward (* = Phosphorothioates S-linkage)	5′ OH-CAAGCAGAAGACGGCATACGAGATGGACGGGTGACTGGAGTTCAGACGTGTGC*T-OH 3′
PCR primer index 18 forward (* = Phosphorothioates S-linkage)	5′ OH-CAAGCAGAAGACGGCATACGAGATGCGGACGTGACTGGAGTTCAGACGTGTGC*T-OH 3′
PCR primer index 19 forward (* = Phosphorothioates S-linkage)	5′ OH-CAAGCAGAAGACGGCATACGAGATTTTCACGTGACTGGAGTTCAGACGTGTGC*T-OH 3′

2.4 Immuno-precipitation

1. Dynabeads Protein A/G beads (Life Technologies) (*see* **Note 1**).
2. Magnetic tube rack (e.g., Life Technologies).
3. Antibody targeting your protein of interest (e.g., anti-AR N-20 Santa Cruz Biotechnology).
4. Nonspecific control IgG (e.g., rabbit IgG).
5. PBS supplemented with 0.5% bovine serum albumin (BSA).

2.5 ChIP-exo

1. RIPA wash buffer: 50 mM Hepes (pH 7.6), 1 mM EDTA, 0.5 M LiCl, 1% Igepal (NP-40), 0.7% sodium deoxycholate.
2. 10 mM Tris–HCl pH 8.0.
3. NEB2 (NEB, supplied with RecJf exonuclease, *see* below).
4. ATP (NEB).
5. Deoxynucleotide (dNTP) solution mix (NEB).
6. NEBNext® End Repair Module (NEB).
7. Klenow fragment (NEB).
8. Thermomixer.
9. Annealing buffer: 10 mM Tris–HCl pH 8.0, 50 mM NaCl, 1 mM EDTA.
10. Annealed P7 exo-adapter forward and reverse oligos (*see* **Note 2**).
11. T4 DNA ligase (NEB) 2000 U/μl.
12. Phi29 DNA polymerase (NEB).
13. In-house Phi29 DNA polymerase buffer: 50 mM Tris–HCl pH 7.5, 10 mM $MgCl_2$, 10 mM $(NH_4)SO_4$, 1 mM DTT. Filter, aliquot, and store at −20 °C.
14. Lambda exonuclease (NEB).
15. RecJf (NEB).

2.6 DNA Elution and Cleanup

1. ChIP elution buffer: 50 mM Tris–HCl, pH 8, 10 mM EDTA, 1% SDS.
2. Proteinase K (20 μg/μl).
3. TE buffer: 50 mM Tris–HCl pH 8, 10 mM EDTA, 1% SDS.
4. Phenol–chloroform–isoamyl alcohol mixture (Sigma).
5. Phase Lock Gel Heavy tubes (5 prime).
6. 5 M NaCl.
7. Glycogen (Life technologies) or suitable carrier for precipitation.
8. Absolute ethanol.
9. 75% Ethanol.
10. 10 mM Tris–HCl pH 8.0.

2.7 ChIP-exo Amplification

1. P7 primer.
2. AMPure XP beads (Beckman Coulter).
3. Annealed P5 exo-adapter forward and reverse oligos (*see* **Note 2**).
4. NEBNext High-Fidelity 2× PCR Master Mix (NEB).
5. PCR Primer mix 50 μM each (1:1 mix of forward and reverse).
6. Certified Low Range Ultra Agarose (Biorad).
7. Dedicated electrophoresis equipment.
8. TAE buffer (Sigma).
9. Ethidium bromide solution (Sigma).
10. NEB Low Molecular Weight DNA Ladder (NEB).
11. Transilluminator.
12. Scalpels.
13. MinElute Gel Extraction Kit (Qiagen).
14. Absolute isopropanol.

2.8 Quality Control

1. Oligonucleotide primers for genomic regions of interest.
2. SYBR green master mix (e.g., Applied Biosystems).
3. PCR plates.
4. Qubit system and dsDNA HS Assay (Life Technologies).
5. Bioanalyzer system and DNA1000 or High Sensitivity DNA Kit (Agilent).

3 Methods

3.1 Cell Culture

1. Maintain LNCaP cells in RPMI medium supplemented with 10% FBS in cell culture incubators (5% CO_2 at 37 °C) and passage at a dilution of 1:3 when approaching confluence with trypsin/EDTA.
2. For ChIP-exo assays involving hormone starvation, split cells and spin down at 300 × *g* for 5 min at RT. Subsequently, gently wash pellet with PBS and resuspend in phenol red-free RPMI supplemented with 10% CSS. Seed 6×10^6 per 15 cm dishes and use two 15 cm dishes per ChIP-exo reaction/antibody; for example, AR and IgG equals four plates in total.

3.2 ChIP-exo

3.2.1 Day 1: Overnight Preparation of Bead-Antibody Complexes

1. 48 h after seeding in CSS medium, wash 50-100 μl Dynabeads (protein A/G depending on antibody species, *see* **Note 1**) per reaction in 2 ml Eppendorf tubes three times with 1–2 ml 0.5% BSA in PBS. Use a magnet to separate the magnetic beads in between washes. Incubate overnight with 5–10 μg of specific antibody or IgG control in 300 μl 0.5% BSA in PBS.

We typically use 10 μg antibody and 50 μl Dynabeads per reaction but different antibodies might require different amounts.

3.2.2 Day 2: Cross-Linking, Harvesting, Sonication, and Immunoprecipitation

1. 72 h after seeding (cells should be 70%+confluent), replace cell culture media with CSS medium supplemented with androgens (e.g., 1 nM DHT or R1881) or an equal volume of ethanol (vehicle) and return cells to the incubator for 4 h. Adjust the preparation of the antibody-bead complexes (day 1) accordingly if longer treatment times are desired.
2. Aspirate medium and replace with 1% formaldehyde in PBS (always prepare fresh). NB: Do not use serum in the cross-linking reaction as this influences the cross-linking efficacy of formaldehyde. Incubate swirling at RT for 10 min (*see* **Note 3**).
3. Quench the cross-linking reaction by adding glycine to a final concentration of 125 mM (1/20 of 2.5 M stock solution). Incubate swirling at RT for 5 min.
4. On ice: Carefully wash plates twice with approximately 15 ml ice-cold PBS. Scrape cells in PBS and transfer to 15 ml Falcon. Spin at 2000×*g* and 4 °C for 5 min and aspirate supernatant.
5. Resuspend pellet in 15 ml LB1 freshly supplemented with proteinase inhibitors and rotate at 4 °C for 10 min to lyse cells. Spin at 2000×*g* and 4 °C for 5 min and aspirate supernatant.
6. Resuspend pellet in 15 ml LB2 freshly supplemented with proteinase inhibitors and rotate at 4 °C for 5 min to wash nuclei. Spin at 2000×*g* and 4 °C for 5 min and aspirate supernatant.
7. Resuspend pellet in 300 μl LB3 freshly supplemented with proteinase inhibitors per 15 cm plate used. Distribute 300 μl each into Eppendorf tubes and rotate at 4 °C for 10 min.
8. Sonicate cells in a Diagenode Bioruptor for approximately 30 cycles of 30 s ON, 30 s OFF to shear chromatin to an average size of 200–300 bp (*see* **Note 4**).
9. Repool lysates and add 1/10 volume of 10% Triton-X in LB3 to quench SDS. Spin at 18,000×*g* for 15 min at 4 °C and transfer supernatant to a new tube.
10. Optional step: Preclearing of chromatin (*see* **Note 5**).
11. If desired, take 25 μl input control and store at −20 °C until tomorrow.
12. Wash the Dynabead-antibody solution from day 1 three times with 0.5% BSA in PBS. After the third wash add identical amounts of sonicated chromatin to every tube with antibody-bead complexes and incubate on a 4 °C rotator overnight.

3.2.3 Day 3: Washing and On-Bead Enzymatic Reactions

1. Using the magnet, wash the bead-antibody-protein–DNA complexes six times with 1–2 ml RIPA wash buffer. Properly resuspend beads between washes. For all subsequent steps,

always prepare enzyme master mixes before washing beads with Tris–HCl pH 8.0. Do not let the beads dry out and do not keep them in Tris–HCl pH 8.0 for an extended period.

2. Set up the end repair master mix and wash the beads twice with 10 mM Tris–HCl pH 8.0. Do not remove tubes from magnet or you will lose material. Make sure that you have thoroughly removed all traces of RIPA or this will inhibit subsequent enzyme reactions. Remove all traces of Tris–HCl pH 8.0 prior to resuspending the beads in the enzyme mix.
3. Blunt DNA fragments using the following end repair reaction. Add 100 μl directly to the beads. Incubate at 30 °C for 30 min on a Thermomixer at 900 rpm. NEB2 is used to maintain a 1 mM DTT concentration (*see* **Note 6**).

   10 μl 10× NEB2 buffer.

   10 μl ATP 10 mM, final 1 mM.

   1 μl dNTPs 10 mM, final 100 μM.

   5 μl End repair enzyme mix (NEB).

   1 μl Klenow fragment (NEB) 5 U/μl.

   73 μl H_2O.

   Total 100 μl.
4. An Illumina-compatible P7 adapter is ligated to the blunted DNA ends.

   Set up a master mix for the ligation of the P7 adapter and wash the beads twice with RIPA wash buffer and 10 mM Tris–HCl pH 8.0 as explained above. Resuspend beads in 100 μl reaction mix and incubate at 25 °C for 60 min on a Thermomixer at 900 rpm.

   10 μl 10× NEB2 buffer.

   10 μl ATP 10 mM, final 1 mM.

   15 μl P7 adapter mix (*see* **Note 6**).

   1 μl T4 DNA ligase (NEB) 2000 U/μl.

   64 μl H_2O.

   Total 100 μl.
5. Ligation of the P7 adapter leaves behind a nick, which needs to be closed.

   Set up a master mix for the nick repair and wash the beads twice with RIPA and 10 mM Tris–HCl pH 8.0 as explained above. Resuspend beads in 100 μl reaction mix and incubate at 30 °C for 20 min on a Thermomixer at 900 rpm.

   1.5 μl Phi29 DNA polymerase (NEB) 10 U/μl, 15 U final.

   1.5 μl dNTPs 10 mM, final 150 μM.

   10 μl 10× homemade Phi29 DNA polymerase buffer.

87 μl H_2O.

Total 100 μl.

The in-house buffer is used to maintain a 1 mM DTT concentration (*see* **Note 6**).

6. Lambda exonuclease is used to digest dsDNA in the 5′–3′ direction until it reaches protected DNA (Fig. 2).

   Set up a master mix for the Lambda exonuclease digestion and wash the beads twice with RIPA and 10 mM Tris–HCl pH 8.0 as explained above. Resuspend beads in 100 μl reaction mix and incubate at 37 °C for 30 min on a Thermomixer at 900 rpm.

   2 μl Lambda exonuclease (NEB) 5 U/μl, 10 U final.

   10 μl 10× Lambda exonuclease buffer.

   88 μl H_2O.

   Total 100 μl.

7. RecJf exonuclease digests ssDNA in the 5′–3′ direction. This reaction is used to reduce noise.

   Set up a master mix for the RecJf exonuclease digestion and wash the beads twice with RIPA and 10 mM Tris–HCl pH 8.0 as explained above. Resuspend beads in 100 μl reaction mix and incubate at 37 °C for 30 min on a Thermomixer at 900 rpm.

   1 μl RecJf (NEB) 30 U/μl.

   10 μl 10× NEB2 buffer.

   89 μl H_2O.

   Total 100 μl.

8. Repeat washes with RIPA and Tris-HCl, remove supernatant and add 200 μl ChIP elution buffer and 5 μl Proteinase K (20 μg/μl). Also include input you froze at −20 °C. Add 175 μl ChIP elution buffer and 5 μl Proteinase K to 25 μl input. Incubate overnight at 65 °C on a Thermomixer at 900 rpm.

3.2.4 Day 4/5: DNA Extraction, Final Enzymatic Reactions, and Size Selection

1. Phenol–chloroform–isoamyl alcohol mixture (PCI) and Phase Lock Gel Heavy tubes are used to clean up DNA to minimize losses.

   Spin Phase Lock tubes at 14,000 × *g* for 1 min to pellet gel. Add 200 μl TE buffer to every sample and mix. Add an equal volume of PCI solution (400 μl) and vortex thoroughly. Transfer to 2 ml Phase Lock tube and spin at 14,000 × *g* for 5 min at RT.

2. Transfer upper phase (approximately 400 μl) to a new Eppendorf tube and add 16 μl 5 M NaCl (200 mM final) and 1 μl glycogen (20 μg/μl), and vortex briefly. Add 1 ml EtOH abs. (ice cold) and store at −80 °C for at least 30 min to facilitate precipitation.

3. Spin at 14,000 × *g* for 15 min at 4 °C and discard supernatant.
4. Wash pellet with ice-cold 75% EtOH and spin again at 14,000 × *g* at 4 °C for 15 min.
   Remove supernatant and let pellet air-dry for 10 min.
5. Elute pellet in 20 μl 10 mM Tris–HCl pH 8.0, incubate at 50 °C for 10 min, and then transfer to clean PCR tubes.
6. Denature DNA at 95 °C for 5 min and put immediately on ice to avoid renaturing of DNA.

   During the 5 min, set up a master mix with the following components. Since DTT concentration does not matter anymore, commercial buffer can be used.

   5 μl P7 primer, 1 μM stock (= 1:100 of 100 μM).

   5 μl Commercial Phi29 DNA polymerase buffer.

   20 μl H_2O.

   Total 30 μl.

   Add to the denatured DNA on ice and incubate at 65 °C for 5 min to anneal, and then cool down to 30 °C for 2 min.
7. Add to every well:

   1 μl Phi29 DNA polymerase (NEB) 10 U/μl.

   1 μl dNTPs, 10 mM.

   Incubate at 30 °C for 20 min followed by 65 °C for 10 min.
8. Briefly spin reaction, add 80 μl of room-temperature AMPure XP beads (1.6 × volumes), and incubate at RT for 15 min. Place tubes on magnet and allow clearing for 15 min. Remove and discard supernatant and wash beads twice with 80% EtOH. Remove all traces of EtOH and let beads air-dry for 5–10 min. Always prepare fresh EtOH every day; you do not want to lose any material due to lower concentrated EtOH and accidentally eluted DNA. Do not remove beads from magnet during washes or you will lose too much material.
9. Resuspend beads in 20 μl RSB and wait for 2 min. Place Eppendorf tubes in magnet, wait for 5 min, and transfer 20 μl supernatant to new PCR tubes.
10. An Illumina-compatible P5 adapter is ligated to the DNA ends. Set up the following master mix for the ligation of the P5 adapter and add 30 μl to each sample.

    5 μl T4 DNA ligase buffer 10×.

    1.5 μl P5 adapter mix.

    1 μl T4 DNA ligase (NEB) 2000 U/μl, final 2000 U.

    22.5 μl H_2O.

    Total 30 μl.

    Incubate at 25 °C for 60 min followed by 65 °C for 10 min.

11. PCR amplify DNA fragments using specific primers containing Illumina-compatible index sequences ("barcodes") that allow multiplexing, i.e., simultaneous sequencing of several samples in one lane (*see* **Note 7**).
    Set up a master mix with the following components:
    25 μl NEBNext High-Fidelity 2× PCR Master Mix (NEB).
    0.5–1 μl Primer mix 50 μM each (1:1 mix of index-specific forward and universal reverse primer, Table 2).
    4–4.5 μl H_2O.
    Total 50 μl.
    Run the following PCR program:
    98 °C for 30 s.
    10–18 cycles of
    98 °C 10 s.
    65 °C 30 s.
    72 °C 30 s.
    72 °C 5 min.
    4 °C forever.
12. Briefly spin PCR reaction, add 50 μl of room-temperature AMPure XP beads (1 vol), and incubate at RT for 15 min. Then place tubes on magnet and allow clearing for 15 min. Remove and discard supernatant and wash beads twice with 80% EtOH. Remove all traces of EtOH and let beads air-dry for 5–10 min.
13. Resuspend beads in 25 μl RSB and wait for 2 min. Place Eppendorf tubes in magnet, wait for 5 min, and transfer 25 μl supernatant to new PCR tubes.
14. Gel electrophoresis is used to isolate fragments of appropriate size.
    Prepare 2% agarose gel (2 g agarose per 100 ml 1× TAE and add appropriate amounts of ethidium bromide (EtBr) or another DNA dye) (*see* **Note 8**).
    To prepare ladder: 8 μl NEB low-molecular-weight DNA Ladder + 3 μl 50% glycerol in 1× TAE, load 11 μl per lane.
    For samples: Add 10 μl 50% glycerol in 1× TAE to 25 μl ChIP-exo sample and load everything into one well.
    Run gel at 120 V for ~40 min.
15. In a darkroom: Using a clean scalpel for every sample, excise bands between 200 and 300 bp and transfer to a 2 ml Eppendorf tube. Stick to a 400 mg maximum and avoid excess transfer of gel that does not contain any DNA. The DNA should be visible and thus easy to cut. Take a picture of the gel before and after excision to document excised range (*see* **Note 9**).

16. DNA is extracted from the gel using the Qiagen MinElute gel extraction kit.
    (a) Weigh gel slice and add three volumes of buffer QG (e.g., 300 μl to 100 mg of gel). Incubate at RT on a rotator/shaker for 10 min to dissolve gel.
    (b) Add 1 gel volume of isopropanol, mix, and transfer to 2 ml column (max. loading capacity = 750 μl, spin and reload if necessary).
    (c) Spin for 1 min at 14,000 × *g*, and discard flow through.
    (d) Add 500 μl QG.
    (e) Spin for 1 min at 14,000 × *g*, and discard flow through.
    (f) Add 750 μl PE and allow to stand for 3 min at RT.
    (g) Spin for 1 min at 14,000 × *g*, and discard flow through.
    (h) Spin for 1 min at 14,000 × *g* to dry column.
    (i) Transfer column to a new Eppendorf tube, add 20 μl pre-warmed (50 °C) EB, and incubate at RT for 1 min.
    (j) Spin for 1 min at 14,000 × *g*.

    This is now a ready-to-be-sequenced Illumina library and can be stored at −20 °C for at least 6 months. The library is sequenced with single-end reads from the ligated P5 adapter and aligned to a reference genome prior to peak calling using a typical algorithm, such as model-based analysis of ChIP-seq (MACS) [17].
17. If sequencing of input samples is desired, prepare input libraries using a conventional protocol, such as the Illumina TruSeq ChIP Sample Preparation Kit (Illumina, IP-202–1012).
18. Prior to sequencing, check DNA concentration and fragment distribution using the Qubit system and Agilent's Bioanalyzer and pool samples if desired (*see* **Note 10**).

    You can use a diluted library (1:10–1:100, depending on concentration) for analysis of enrichment using quantitative real-time PCR with oligonucleotides to your genomic region of interest. Calculate the fold enrichment over IgG using the 2ΔΔCt method.

4 Notes

1. Dynabeads Protein A and G beads exhibit species-dependent differences in affinities towards antibodies. Consult Life Technologies' homepage when planning your experiment.
2. To prepare annealed dsDNA P7 adapter mix, mix equal volumes of 100 μM primer stocks P7 FWD and P7 REV (e.g., 100 μl of both) (Table 2) with four volumes (e.g., 800 μl) of

annealing buffer (10 mM Tris–HCl pH 8.0, 50 mM NaCl, 1 mM EDTA). Heat mixture to 95 °C for 5 min and then move it to room temperature for 2 h. The mixture will gradually cool down and allow the oligos to anneal. This will yield dsDNA adapters used in this reaction. Aliquot and store at −20 °C until ready to use. Annealed P5 exo-adapters are prepared in the same manner using the forward and reverse primers listed in Table 2.

3. Use this as a guideline only. Formaldehyde concentration and cross-linking duration might require optimization if ChIPped for other transcription factors. Using other cross-linking agents, such as imidoesters or NHD-esters, might be possible/ necessary/applicable for your specific experiment. In addition, performing native ChIP, i.e., using no cross-linking agent, can be possible but is generally only suitable for proteins that are very tightly attached to DNA.
4. The optimal sonication time and intensity depend on various factors, such as your sonication device, your cell line, and your transcription factor of interest. Thus, this step requires optimization and optimal sonication time should be determined prior to starting the experiment to ensure optimal resolution. To assess shearing efficacy, take an aliquot (e.g., 20 μl), reverse cross-link, and clean up DNA according to the steps mentioned under Subheading 3.2.3, **step 8**. Subsequently, load varying amounts onto a 2% agarose gel and visualize under UV light.
5. If you experience high background in your reactions, it might be helpful to preclear the chromatin. To perform this, incubate it with 20 μl Dynabeads for 1 h at 4 °C rotating and transfer the supernatant to a clean tube using the magnet prior to proceeding with the next step. This will clear your chromatin of fragments that bind unspecifically to the Dynabeads.
6. Do not use the NEB end repair buffer that comes with the mix. It contains 10 mM DTT and will elute material. NEB2 contains only 1 mM DTT and is also suitable for this reaction. The same applies to the Phi29 reactions where the commercial buffer contains 10 mM DTT and a custom-made buffer is used instead.
7. When multiplexing (simultaneous sequencing of multiple samples in a single lane), follow Illumina's Adapter Tube Pooling Guidelines (refer to Table 3 for pooling strategies).

   In general, ChIP-exo allows higher plexity than ChIP-seq and Serandour et al. successfully sequenced and demultiplexed up to 12 samples, each with at least 15 million reads on an Illumina HiSeq machine.

   Both the optimal primer concentration and the amount of PCR cycles depend on the amount of ChIPped material. Thus,

Table 3
Illumina's adapter tube pooling guidelines

Plexity	Option	Adapters
2	1	Index 6 and index 12
	2	Index 5 and index 19
3	1	Index 2, index 7, and index 19
	2	Index 5, index 6, and index 15
	3	2-plex option with any other adapter
4	1	Index 5, index 6, index 12, and index 19
	2	Index 2, index 4, index 7, and index 16
	3	3-plex option with any other adapter
5 or higher	1	4-plex option with any other adapter

this step might require optimization in order to avoid too little material or overamplification, which might lead to large amounts of duplicate reads and underrepresented libraries during sequencing.

8. When loading the gel, always leave an empty well between every sample/marker to avoid cross-contamination. When using other DNA dyes than ethidium bromide, such as SYBR gold, make sure to do a test run since many dyes are known to influence DNA migration.
9. The excision range determines the libraries' resolutions. If lower resolutions are required, cut at higher range and vice versa. Do not cut below 200 bp to avoid contamination with unconjugated primers.
10. When multiplexing (simultaneous sequencing of multiple samples in a single lane), it is essential that sample concentrations are roughly equal to ensure a similar sequencing depth for all samples. Thus, calculate mean fragment sizes using Agilent's Bioanalyzer and concentrations with the Qubit system (Life Technologies). Subsequently, calculate the molarity of every sample and adjust them according to your sequencing facility's instructions.

References

1. Solomon MJ, Larsen PL, Varshavsky A (1988) Mapping protein–DNA interactions in vivo with formaldehyde: evidence that histone H4 is retained on a highly transcribed gene. Cell 53(6):937–947
2. Ren B, Robert F, Wyrick JJ, Aparicio O, Jennings EG, Simon I, Zeitlinger J, Schreiber J, Hannett N, Kanin E, Volkert TL, Wilson CJ, Bell SP, Young RA (2000) Genome-wide location and function of DNA binding proteins.

Science 290(5500):2306–2309. doi:10.1126/science.290.5500.2306

3. Barski A, Cuddapah S, Cui K, Roh T-Y, Schones DE, Wang Z, Wei G, Chepelev I, Zhao K (2007) High-resolution profiling of histone methylations in the human genome. Cell 129(4):823–837. doi:10.1016/j.cell.2007.05.009
4. Robertson G, Hirst M, Bainbridge M, Bilenky M, Zhao Y, Zeng T, Euskirchen G, Bernier B, Varhol R, Delaney A, Thiessen N, Griffith OL, He A, Marra M, Snyder M, Jones S (2007) Genome-wide profiles of STAT1 DNA association using chromatin immunoprecipitation and massively parallel sequencing. Nat Methods 4(8):651–657. doi:10.1038/nmeth1068
5. Landt SG, Marinov GK, Kundaje A, Kheradpour P, Pauli F, Batzoglou S, Bernstein BE, Bickel P, Brown JB, Cayting P, Chen Y, DeSalvo G, Epstein C, Fisher-Aylor KI, Euskirchen G, Gerstein M, Gertz J, Hartemink AJ, Hoffman MM, Iyer VR, Jung YL, Karmakar S, Kellis M, Kharchenko PV, Li Q, Liu T, Liu XS, Ma L, Milosavljevic A, Myers RM, Park PJ, Pazin MJ, Perry MD, Raha D, Reddy TE, Rozowsky J, Shoresh N, Sidow A, Slattery M, Stamatoyannopoulos JA, Tolstorukov MY, White KP, Xi S, Farnham PJ, Lieb JD, Wold BJ, Snyder M (2012) ChIP-seq guidelines and practices of the ENCODE and modENCODE consortia. Genome Res 22(9):1813–1831. doi:10.1101/gr.136184.111
6. Shankaranarayanan P, Mendoza-Parra M-A, Walia M, Wang L, Li N, Trindade LM, Gronemeyer H (2011) Single-tube linear DNA amplification (LinDA) for robust ChIP-seq. Nat Methods 8(7):565–567. doi:10.1038/nmeth.1626
7. Adli M, Bernstein BE (2011) Whole-genome chromatin profiling from limited numbers of cells using nano-ChIP-seq. Nat Protoc 6(10):1656–1668. doi:10.1038/nprot.2011.402
8. Goren A, Ozsolak F, Shoresh N, Ku M, Adli M, Hart C, Gymrek M, Zuk O, Regev A, Milos PM, Bernstein BE (2010) Chromatin profiling by directly sequencing small quantities of immunoprecipitated DNA. Nat Methods 7(1):47–49. doi:10.1038/nmeth.1404
9. Fanelli M, Amatori S, Barozzi I, Minucci S (2011) Chromatin immunoprecipitation and high-throughput sequencing from paraffin-embedded pathology tissue. Nat Protoc 6(12):1905–1919. doi:10.1038/nprot.2011.406
10. Zwart W, Koornstra R, Wesseling J, Rutgers E, Linn S, Carroll JS (2013) A carrier-assisted ChIP-seq method for estrogen receptor-chromatin interactions from breast cancer core needle biopsy samples. BMC Genomics 14:232. doi:10.1186/1471-2164-14-232
11. Rhee HS, Pugh BF (2011) Comprehensive genome-wide protein–DNA interactions detected at single-nucleotide resolution. Cell 147(6):1408–1419. doi:10.1016/j.cell.2011.11.013
12. Serandour AA, Brown GD, Cohen JD, Carroll JS (2013) Development of an illumina-based ChIP-exonuclease method provides insight into FoxA1-DNA binding properties. Genome Biol 14(12):R147. doi:10.1186/gb-2013-14-12-r147
13. Massie CE, Lynch A, Ramos-Montoya A, Boren J, Stark R, Fazli L, Warren A, Scott H, Madhu B, Sharma N, Bon H, Zecchini V, Smith D-M, Denicola GM, Mathews N, Osborne M, Hadfield J, Macarthur S, Adryan B, Lyons SK, Brindle KM, Griffiths J, Gleave ME, Rennie PS, Neal DE, Mills IG (2011) The androgen receptor fuels prostate cancer by regulating central metabolism and biosynthesis. EMBO J 30(13):2719–2733. doi:10.1038/emboj.2011.158
14. Sharma NL, Massie CE, Ramos-Montoya A, Zecchini V, Scott HE, Lamb AD, MacArthur S, Stark R, Warren AY, Mills IG, Neal DE (2013) The androgen receptor induces a distinct transcriptional program in castration-resistant prostate cancer in man. Cancer Cell 23(1):35–47. doi:10.1016/j.ccr.2012.11.010
15. Yu J, Yu J, Mani R-S, Cao Q, Brenner CJ, Cao X, Wang X, Wu L, Li J, Hu M, Gong Y, Cheng H, Laxman B, Vellaichamy A, Shankar S, Li Y, Dhanasekaran SM, Morey R, Barrette T, Lonigro RJ, Tomlins SA, Varambally S, Qin ZS, Chinnaiyan AM (2010) An integrated network of androgen receptor, polycomb, and TMPRSS2-ERG gene fusions in prostate cancer progression. Cancer Cell 17(5):443–454. doi:10.1016/j.ccr.2010.03.018
16. Sahu B, Laakso M, Ovaska K, Mirtti T, Lundin J, Rannikko A, Sankila A, Turunen J-P, Lundin M, Konsti J, Vesterinen T, Nordling S, Kallioniemi O, Hautaniemi S, Jänne OA (2011) Dual role of FoxA1 in androgen receptor binding to chromatin, androgen signalling and prostate cancer. EMBO J 30(19):3962–3976. doi:10.1038/emboj.2011.328
17. Zhang Y, Liu T, Meyer CA, Eeckhoute J, Johnson DS, Bernstein BE, Nusbaum C, Myers RM, Brown M, Li W, Liu XS (2008) Model-based analysis of ChIP-Seq (MACS). Genome Biol 9(9):R137. doi:10.1186/gb-2008-9-9-r137

Chapter 9

Methods to Identify Chromatin-Bound Protein Complexes: From Genome-Wide to Locus-Specific Approaches

Charles E. Massie

Abstract

High-throughput sequencing approaches coupled with functional genomics experiments have facilitated a rapid growth in our understanding of chromatin biology, from genome-wide maps of transcription factor binding and histone modifications to insights into higher order chromatin organization under specific cellular conditions. However in most cases these methods require a prior knowledge of the system of interest (e.g., targets for immunoprecipitation or modulation) and therefore are limited in their utility to identify novel components of pathways or for the study of uncharacterized pathways. Several orthologous proteomics approaches have been developed recently that bridge this gap, allowing the identification of protein complexes globally or at specific genomic loci. In this chapter the relative advantages of each approach will be explored and a detailed protocol given for DNA pull-down of a specific androgen receptor (AR) genomic target.

Key words Transcriptional regulation, Transcription, Chromatin, Proteomics, Genomics, High-throughput screen, Androgen receptor (AR), Nuclear hormone receptor, Prostate cancer

1 Introduction

Mechanisms that regulate gene expression and chromatin organization are central to all cellular processes. As such the upstream signaling pathways and downstream effectors that regulate these processes are fundamentally important in development and disease biology. Over recent decades methods to map genome-proteome interactions [1–4] and to identify higher order chromatin structures [5, 6] have given new insights into the mechanisms that underlie these central cellular pathways. These methods have provided important insights into chromatin regulation and organization in a number of biological systems. However, these methods (e.g., chromatin immunoprecipitation, ChIP) are most useful in mapping the genomic landscape of known transcriptional regulators and can only identify co-enrichment of other transcriptional regulators through DNA sequence motif analysis [7–9], which

Iain J. McEwan (ed.), *The Nuclear Receptor Superfamily: Methods and Protocols*, Methods in Molecular Biology, vol. 1443,
DOI 10.1007/978-1-4939-3724-0_9,

again is limited to well-studied, known regulators and direct DNA-binding proteins. Therefore, additional methods are required to more comprehensively identify components of chromatin-bound protein complexes and to identify regulators at specific genomic loci. This shortcoming is highlighted by the identification of chromatin-bound proteins which could not have been predicted based on prior knowledge, nor by DNA sequence analysis (e.g., endocytic adaptors, kinases with cytoplasmic roles, and transmembrane receptors [10–13]).

Over recent years improvements in mass spectroscopy have facilitated the development of methods to bridge this experimental gap. These methods include extensions of the ChIP methodology to allow the identification of co-enriched proteins and also methods to enrich proteins bound to specific genomic loci (summarized in Table 1).

Table 1
Comparison of existing methods that can be used to identify protein-DNA complexes

Method	Starting material	Scope of assay	Specific advantages	Specific disadvantages	References
RIME	~10^7 cells	Genome-wide	Endogenous complexes and target proteins, rapid assay	Averages signal across all complexes, antibody limitations	[14, 15]
Tag-ChIP	~10^{6-7} cells	Genome-wide	Efficient enrichment, no antibody limitations	Require stable cell lines for exogenous/endogenous tagged protein	[16, 17]
LexA/TAL4-binding site PD	~10^{11-12} cells	Locus specific	Endogenous locus analysis (rather than specific protein complex)	Require stable cell lines with binding site knock-in, possible effects on locus function	[18, 19]
TAL/CRISPR-ChAP	~10^{10-11} cells	Locus specific	Endogenous locus analysis	Likely low specificity (untested), possibly affecting endogenous protein complexes	[20, 21]
PICh	~10^{11} cells	Locus specific	Endogenous locus analysis	Large amount of starting material	[22, 23]
DNA-PD	~10^{6-7} cells	Locus specific	Locus/motif assay, low starting material, rapid	In vitro method, requires reconstitution of complexes	[24–29]

RIME rapid immunoprecipitation mass spectrometry of endogenous proteins, *ChIP* chromatin immunoprecipitation, *ChAP* chromatin affinity purification, *PD* pull-down

1.1 Global Profiles of Protein Complexes

Antibody and tag affinity enrichment methods based on ChIP include rapid immunoprecipitation mass spectrometry of endogenous proteins (RIME) that enriches endogenous chromatin-bound protein complexes [14, 15] and biotin-streptavidin capture of tagged proteins to enrich chromatin-bound protein complexes [16, 17]. These methods allow a genome-wide view of protein complexes involving a specific "bait" protein and have successfully identified new co-regulators even in well-studied systems [14]. These protein "bait" approaches are most useful to characterize co-regulators of specific proteins, but may be less useful to identify dynamic changes at regulatory elements (e.g., if the "bait" protein cycles on/off chromatin). Having a genome-wide view of protein complexes also means that the signal obtained is averaged across all complexes and therefore biases these methods towards the identification of core components and may not accurately reflect the diversity of protein complexes which have different functional effects at different loci (e.g., divergent complexes involving transcription factors that activate or repress target gene expression at different genomic loci).

1.2 Locus-Specific Proteomics

A number of approaches have been developed to identify proteins that bind to specific genomic loci (*see* Table 1). These include DNA-sequence capture approaches (e.g., PICh [22]), introduction of exogenous "bait" DNA sequences into loci of interest (e.g., LexA-ChAP [18]), using RNA-guided transcription factors (e.g., tagged TAL1 or CRISPR [20, 21]), and finally in vitro DNA pull-down assays to reconstitute protein complexes on specific DNA sequences [24–26]. Locus-specific methods which enrich proteins on endogenous cellular chromatin offer great promise for the de novo identification of proteins and complexes bound at specific genomic elements. Such methods also offer the potential to monitor dynamic changes as transcription factors bind or dissociate. However the very large amounts of starting material required for all such methods ($>10^{10}$–10^{12} cell per reaction) have limited their application to cell types that can be cultured in bioreactors (e.g., yeast or suspension cell lines) and for the most part to high-copy genomic elements. For example the "proteomics of isolated chromatin segments" (PICh) method has been successfully applied to study the telomere proteome in suspension batch-cultured cell lines [22], which was made more feasible by the ~50-fold greater abundance of telomeres compared to single-copy genomic loci.

In contrast methods based on in vitro DNA pull-down require several orders of magnitude fewer cells as input compared to methods that enrich endogenous loci from cellular chromatin (10^6–10^7 cells per reaction). Such DNA pull-down methods are based on the same principle as electromobility shift assays (EMSA) [30] in that they use naked double-stranded DNA molecules as scaffolds on which DNA-binding complexes can be reconstituted from native cell lysates (Fig. 1a) [24, 27, 28]. Therefore their biggest limitation is the extent to which these reconstituted DNA-protein complexes reflect the complexes bound at endogenous genomic loci. Two recently published studies have shown that these methods can be used to reconstitute endogenous complexes, evidenced by the successful identification of estrogen receptor (ERα) and the androgen receptor (AR) protein complexes using in vitro DNA pull-down methods [24, 26]. Therefore, such in vitro DNA pull-down methods may offer a practical screening tool to identify protein complexes and novel protein components using templates from

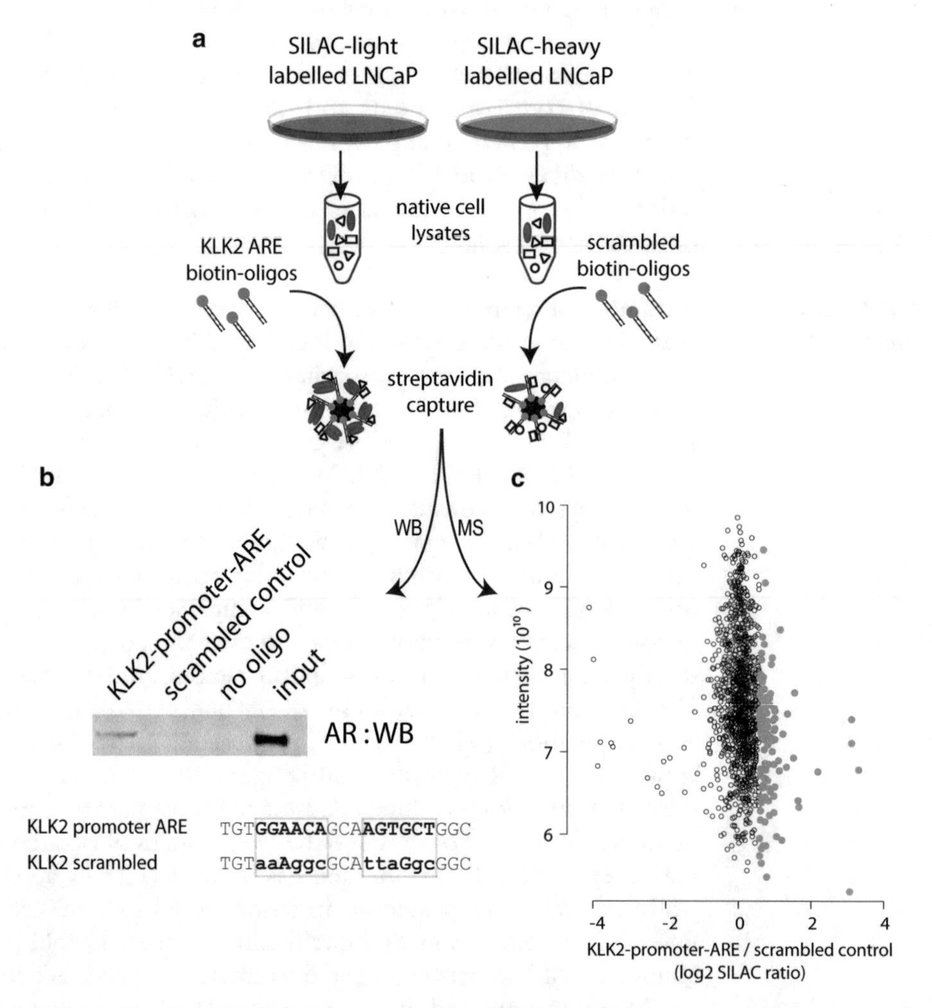

Fig. 1 Overview of in vitro DNA pull-down assay using KLK2 promoter androgen response element (ARE) sequences and LNCaP cell lysates. (**a**) Schematic overview of the DNA pull-down method. (**b**) Example Western blot validation of KLK2 promoter ARE and scrambled control DNA pull-downs (*sequences highlighted below*, *boxes* indicate core motifs and *lowercase* indicates scrambled bases). (**c**) Example mass spectrometry results for SILAC test versus control DNA pull-down assay using the KLK2 promoter ARE. *Filled grey data points* indicate proteins that passed significance testing for enrichment in wild-type versus scrambled control pull-down reactions

specific genomic loci. Candidates identified using these approaches will require validation using additional biochemical methods, ChIP, or functional experiments (e.g., reporter assays or effects on endogenous target loci).

2 Materials

2.1 Cell Growth and Lysis

1. LNCaP cells (*see* **Note 1**) and growth media for cell expansion and initial optimization: RPMI (Invitrogen) supplemented with 10% fetal bovine serum (FBS, HyClone).
2. Stable isotope labeling with amino acids in cell culture (SILAC) medium for metabolic labeling: RPMI (lacking lysine and arginine, Gibco) supplemented with either $^2H_4$-lysine (Sigma Isotec) for heavy-SILCA medium or unlabeled lysine for light-SILAC medium.
3. Modified HKMG buffer: 10 mM Hepes, 100 mM KCl, 5 mM $MgCl_2$, 10% glycerol, 0.5% NP40, 1 mM DTT (added just before use or 0.1% BME), 1× complete protease inhibitor cocktail (Roche). This buffer was supplemented with 10 μM $ZnCl_2$ and 10 pM R1881 for androgen receptor assays.
4. Ice-cold PBS supplemented with complete protease inhibitors (Roche).
5. Cell scrapers (e.g., 89260-222, VWR).
6. Sonicator (e.g., Bioruptor, Diagenode), bench-top centrifuge, and microfuge.

2.2 Preparing Control and Target-Binding Sequences

1. Magnetic tube rack, vortex, tube rotator, microcentrifuge.
2. Magnesphere beads (Z5481, Promega).
3. Biotin-tagged HPLC-purified oligonucleotides for target sequence of interest and scrambled control (Table 2; *see* **Note 2**).
4. Heat block or PCR cycler.

2.3 Pre-clearing and Oligo Pull-Down

1. Magnetic tube rack.
2. Vortex, tube rotator (or rolling mixer), and microcentrifuge.
3. Biotin-tagged oligonucleotides for scrambled control pre-bound to Magnesphere beads (Z5481, Promega) (Table 1, *see* **Note 2**).
4. 1× Denaturing sample loading buffer (e.g., LDS sample buffer, 84788 Fisher Scientific supplemented with 2-mercapto-ethanol or DTT) (*see* **Note 4**).
5. Precise Tris-HEPES gels, 10×8.5 cm (25204, Thermo Scientific).
6. Colloidal Coomassie stain (LC6025, Life Technologies).

Table 2
Test and scrambled control oligonucleotides selected from the KLK2 promoter AR-binding site

Oligonucleotide sequence	Descriptive name
Biotin-TGT**GGAACA**GCAAGTGCTGGC	KLK2-promoter-ARE-S
Biotin-GCC**AGCACT**TGC**TGTTCC**ACA	KLK2-promoter-ARE-AS
Biotin-TGTaaAggcGCAttaGgcGGC	Scrambled-KLK2-promoter-ARE-S
Biotin-GCCgcCtaaTGCgccTttACA	Scrambled-KLK2-promoter-ARE-AS

The target sequence was selected using ChIP-seq enrichment analysis and AR motif analysis to identify a core AR-binding site (*see* **Note 3**). Core AR binding motifs are highlighted in bold. Scrambled bases shown in lowercase. Oligonucleotides were modified with a 5′ biotin tag

3 Methods

3.1 Cell Culture and Harvesting Cell Lysates

1a. For initial optimization by DNA pull-down and Western blot analysis: LNCaP cells grown in RPMI supplemented with 10% FBS to ~70% confluence (1× T75cm^2 flask per pull-down, equivalent to ~5×10^6 cells). At least two pull-down reactions will be required for each target sequence (i.e., for test and scrambled control).

1b. For quantitative mass spectrometry protocol LNCaP cells grown in heavy or light SILAC media for three passages and then grown to ~70% confluence (1× T75cm^2 flask per pull-down, equivalent to ~5×10^6 cells).

2. Wash cells with ice-cold 1× PBS and harvest on ice with a cell scraper.
3. Pellet cells at 1500 × *g* for 3 min at 4 °C. Resuspend cells in 1 ml modified HKMG buffer per pull-down (*see* **Note 5**).
4. Sonicate cell suspensions in a pre-chilled water bath sonicator for 5 min at full power (or in ice water with a probe sonicator, after defining the optimal sonication conditions to liberate nuclear proteins and disrupt chromatin—*see* **Note 6**).
5. Centrifuge samples at 13,000 × *g* for 10 min at 4 °C. Transfer supernatants to a fresh tube and store on ice (*see* **Note 7**).

3.2 Annealing Complementary Biotinylated Oligonucleotides

1. For each pair of complementary oligonucleotides mix 30 μl sense oligo (2 μg/μl), 30 μl antisense oligo (2 μg/μl), and 6.6 μl 10× PCR buffer (or 100 mM Tris pH 8, 15 mM $MgCl_2$).
2. Incubate at 100 °C on a heat block for 5 min, then remove block from heater, and allow tubes to cool slowly to ambient temperature in the block. Alternatively, incubate samples on a thermocycler: 100 °C 5 min and then −1 °C every 5 s for 70 cycles.

3.3 Binding Double-Stranded Oligonucleotides to Magnetic Beads

1. Use one tube of Magnesphere beads for each DNA pull-down reaction. Immobilize beads on a magnetic tube rack and wash three times with 1 ml HKMG buffer (removing the magnet each time to allow thorough washing of the beads).
2. Resuspend beads in 100 μl HKMG buffer and take 50 μl to a fresh tube to bind the pre-annealed control oligonucleotides (used for pre-clearing each DNA pull-down reaction). Take the remaining 50 μl of bead slurry to a separate tube to bind pre-annealed oligonucleotides for the target sequence of interest (i.e., test or scrambled control sequences).
3. Add 10 μl of pre-annealed oligonucleotides (10 μg) to the magnetic bead slurry and incubate at room temperature for 15 min with agitation (e.g., on tube rotator).
4. Immobilize beads on a magnetic tube rack and wash three times with 1 ml HKMG buffer. Resuspend bead-oligonucleotide complexes in 50 μl HKMG buffer and store on ice.

3.4 Pre-clearing and Oligonucleotide Pull-Down

1. Pre-clear each 1ml native cell lysate (from **step 3.1.5**) using 50 μl of bead-oligo complexes loaded with control sequences (e.g., scrambled control probes lacking the binding site of interest, *see* **Note 2**). Incubate for 1 h at 4 °C on a rolling mixer (e.g., in cold room).
2. Immobilize beads on a magnetic tube rack and transfer the pre-cleared supernatant to a fresh tube.
3. For test DNA pull-down reactions add 50 μl of bead-bound target sequence to 1 ml pre-cleared lysates from **step 3.4.2** (e.g., using lysates from light isotope-labeled LNCaP cells if processing for mass spectrometry—*see* **Note 8**). For control DNA pull-down reactions add 50 μl of bead-bound scrambled control sequence to 1ml pre-cleared lysates from **step 3.4.2** (e.g., using heavy isotope-labeled LNCaP lysates if processing for MS). Incubate for 4–16 h at 4 °C on a rolling mixer (e.g., in cold room).
4. Immobilize bead-oligo-protein complexes on a magnetic tube rack, discard the supernatant, and wash beads five times with ice-cold modified HKMG buffer.

5a. Elution of bound proteins for Western blot analysis (*see* **Note 4**): Resuspend beads in 100 μl 1× denaturing sample loading buffer. Boil samples for 3 min at 100 °C, capture beads on a magnetic rack, load 20 μl of the DNA pull-down sample into each well of a denaturing polyacrylamide gel (e.g., Tris-HEPES 4–20% gradient gel), and process for Western blotting to detect the target protein and negative control proteins or to validate candidates identified by mass spectroscopy. Be sure to include test, scrambled control, and total input lanes to allow assessment of specific enrichment and for troubleshooting (*see* Fig. 1b and **Note 9**).

5b. Elution of bound proteins for quantitative mass spectroscopy (*see* **Note 4**): Resuspend test (e.g., heavy isotope labeled) and control (e.g., light isotope labeled) bead-oligo-protein complex slurry in 50 μl modified HKMG buffer and combine in a single tube. Immobilize beads on a magnetic tube rack, discard the supernatant, and resuspend in 30 μl 1× denaturing loading buffer. Boil for 3 min at 100 °C, immobilize beads on a magnetic tube rack, and load all of the supernatant on a single lane of a denaturing polyacrylamide gel, leaving at least one lane gap between any other lanes. Run electrophoresis to separate eluted proteins based on their electrophoretic mobility. Stain gels using colloidal Coomassie and slice each lane into eight equal pieces. Process gel bands according to your local mass spectroscopy protocol (e.g., trypsin in-gel digestion, desalting, and LC-MS—*see* **Note 10** and Fig. 1c). Candidate proteins identified using this practical screening method should be validated using orthologous methods (*see* **Note 11**).

4 Notes

1. Suitable cell lines for oligo pull-down experiments: Ideally cells matching a particular phenotype (i.e., cell type and disease state) with appropriate expression status of the target loci (i.e., corresponding to the DNA-binding sequence selected) should be used in these experiments, to ensure that the relevant transcriptional machinery is expressed in the test cell line. We used LNCaP prostate cancer cells in the example shown, which was the same cell line used to identify the target loci (using ChIP-seq analysis) and expresses the gene associated with the target loci. In some circumstances it may be interesting to compare oligo pull-down from cell types with phenotypic differences (e.g., hormone-dependant and hormone-relapsed prostate cancer).
2. Selection of test and scrambled control oligonucleotides: Regions of interest could be short transcription factor-binding motifs from binding sites identified by ChIP-sequencing (as in the example test case presented here) or could include a longer DNA sequence if it is desirable to identify additional DNA-binding factors which could bind to the DNA probe independently of the core transcription factor of interest. Control sequences should offer a contrast and allow control for non-specific charge-based interactions, so they could either be scrambled controls (as presented here) or use other genomic loci not bound by the factor of interest (e.g., like unbound control regions used in ChIP-qPCR analysis).
3. When using standard ChIP-seq analysis the resultant genomic binding "peaks" are commonly in the order of 100s of bases in

length and therefore a secondary motif enrichment analysis is required to identify core transcription factor-binding sites (using PWM from databases such as JASPAR or alternatively using PWM derived from ChIP-seq data using de novo motif analysis). However, motif analysis would not be necessary if data from the modified ChIP-exo protocol are used as a starting point, where peaks are commonly an order of magnitude shorter (e.g., 25–50 bp), thereby directly identifying core binding sites [3, 4].

4. Elution of enriched proteins can be achieved by multiple methods that offer differing benefits with yield and background contamination. Previous studies have successfully employed either elution using denaturing conditions [31]; mild elution using desthiobiotin-tagged oligos and biotin elution [22]; or introduction of restriction enzyme sites into probe-tag sequences and elution by restriction digest following pull-down [25].

5. The method presented here uses whole-cell lysates; however other methods have used nuclear fractionation [24, 25] which may decrease nonspecific contamination with cytoplasmic proteins for the cellular conditions under investigation.

6. Optimization of sonication should be achieved by identifying the minimal sonication required to solubilize chromatin-bound proteins from cell lysates (i.e., to avoid loss of chromatin-bound proteins in the insoluble cell debris fraction following centrifugation at 13,000 × *g* for 10 min). This may be tested by Western blot analysis of lysate supernatants and cell debris pellets using different sonication conditions.

7. Native lysates should ideally be used for DNA pull-down assays fresh on the same day, but may be snap-frozen in ethanol-dry ice or liquid nitrogen for later use.

8. When using quantitative mass spectroscopy it is essential to include biological replicates and advisable to perform label swap experiments (e.g., heavy-SILAC test vs. light-SILAC control in addition to light-SILAC test vs. heavy-SILAC control experiments).

9. If Western blot analysis shows no specific enrichment of control proteins in test versus control DNA pull-down experiments further optimization will be required. As is the case for the related EMSA method, it is not uncommon for different DNA template sequences or different cell types to require some initial optimization. To provide a good starting point for optimization DNA pull-down assays should be quality controlled by Western blotting for target proteins that bind the endogenous locus, unrelated control proteins (preferably nuclear and DNA bound), and known-interaction partners of

the target protein (or known regulators of the loci or interest). Optimization may be achieved by increasing stringency (e.g., longer or more wash steps, increasing detergents in buffers) or decreasing stringency (e.g., fewer or shorted wash steps, changing salt concentrations), as required. In addition one may consider using an alternative control probe for pre-clearing (*see* **Note 2**), including blocking probes in the target incubations (e.g., biotin-free scrambled control or poly[dI.dC]), using nuclear lysates (*see* **Note 5**) or using an alternative elution method (*see* **Note 4**). Numerous commercial kits are available for optimizing EMSAs and these may provide useful starting points for further optimization.

10. Numerous mass spectrometry platforms exist and may require different workflows. A detailed example of a successful downstream mass spectrometry protocol and data analysis pipeline for DNA pull-downs is given by Mittler et al. [25].
11. Candidate proteins enriched using this method should first be validated by biological replicate DNA pull-down and Western blotting (ideally using different cells, oligonucleotide sequences, or an orthologous design, such as RNAi or treatment contrast). Endogenous or tagged ChIP-qPCR could be used to assess candidate protein recruitment at the endogenous genomic loci. In addition it may be valuable to combine this approach with other proteomic approaches summarized in Table 1 or to undertake functional validation experiments to assess the role that enriched proteins may play in the biological system under investigation (e.g., RNAi knockdown to assess the effects on target gene expression and relevant cellular phenotypes).

Acknowledgements

Mohammad Asim generously provided critical feedback on this chapter. C.E.M. is a research associate funded at the time of writing by a Cancer Research UK Programme Grant.

References

1. Orlando V, Paro R (1993) Mapping Polycomb-repressed domains in the bithorax complex using in vivo formaldehyde cross-linked chromatin. Cell 75(6):1187–1198
2. Robertson G, Hirst M, Bainbridge M, Bilenky M, Zhao Y, Zeng T, Euskirchen G, Bernier B, Varhol R, Delaney A, Thiessen N, Griffith OL, He A, Marra M, Snyder M, Jones S (2007) Genome-wide profiles of STAT1 DNA association using chromatin immunoprecipitation and massively parallel sequencing. Nat Methods 4(8):651–657. doi:10.1038/nmeth1068
3. Rhee HS, Pugh BF (2012) ChIP-exo method for identifying genomic location of DNA-binding proteins with near-single-nucleotide accuracy. Curr Protoc Mol Biol Chapter 21:Unit 21 24. doi:10.1002/0471142727.mb2124s100

4. Serandour AA, Brown GD, Cohen JD, Carroll JS (2013) Development of an Illumina-based ChIP-exonuclease method provides insight into FoxA1-DNA binding properties. Genome Biol 14(12):R147. doi:10.1186/gb-2013-14-12-r147
5. Sexton T, Kurukuti S, Mitchell JA, Umlauf D, Nagano T, Fraser P (2012) Sensitive detection of chromatin coassociations using enhanced chromosome conformation capture on chip. Nat Protoc 7(7):1335–1350. doi:10.1038/nprot.2012.071
6. Gavrilov A, Eivazova E, Priozhkova I, Lipinski M, Razin S, Vassetzky Y (2009) Chromosome conformation capture (from 3C to 5C) and its ChIP-based modification. Methods Mol Biol 567:171–188. doi:10.1007/978-1-60327-414-2_12
7. Mathelier A, Zhao X, Zhang AW, Parcy F, Worsley-Hunt R, Arenillas DJ, Buchman S, Chen CY, Chou A, Ienasescu H, Lim J, Shyr C, Tan G, Zhou M, Lenhard B, Sandelin A, Wasserman WW (2014) JASPAR 2014: an extensively expanded and updated open-access database of transcription factor binding profiles. Nucleic Acids Res 42(Database issue):D142–D147. doi:10.1093/nar/gkt997
8. Ma W, Noble WS, Bailey TL (2014) Motif-based analysis of large nucleotide data sets using MEME-ChIP. Nat Protoc 9(6):1428–1450. doi:10.1038/nprot.2014.083
9. Bailey TL, Elkan C (1994) Fitting a mixture model by expectation maximization to discover motifs in biopolymers. Proc Int Conf Intell Syst Mol Biol 2:28–36
10. Mills IG, Gaughan L, Robson C, Ross T, McCracken S, Kelly J, Neal DE (2005) Huntingtin interacting protein 1 modulates the transcriptional activity of nuclear hormone receptors. J Cell Biol 170(2):191–200. doi:10.1083/jcb.200503106
11. Dawson MA, Bannister AJ, Gottgens B, Foster SD, Bartke T, Green AR, Kouzarides T (2009) JAK2 phosphorylates histone H3Y41 and excludes HP1alpha from chromatin. Nature 461(7265):819–822. doi:10.1038/nature08448
12. Lin SY, Makino K, Xia W, Matin A, Wen Y, Kwong KY, Bourguignon L, Hung MC (2001) Nuclear localization of EGF receptor and its potential new role as a transcription factor. Nat Cell Biol 3(9):802–808. doi:10.1038/ncb0901-802
13. Li W, Park JW, Nuijens A, Sliwkowski MX, Keller GA (1996) Heregulin is rapidly translocated to the nucleus and its transport is correlated with c-myc induction in breast cancer cells. Oncogene 12(11):2473–2477
14. Mohammed H, D'Santos C, Serandour AA, Ali HR, Brown GD, Atkins A, Rueda OM, Holmes KA, Theodorou V, Robinson JL, Zwart W, Saadi A, Ross-Innes CS, Chin SF, Menon S, Stingl J, Palmieri C, Caldas C, Carroll JS (2013) Endogenous purification reveals GREB1 as a key estrogen receptor regulatory factor. Cell Rep 3(2):342–349. doi:10.1016/j.celrep.2013.01.010
15. Ji Z, Mohammed H, Webber A, Ridsdale J, Han N, Carroll JS, Sharrocks AD (2014) The forkhead transcription factor FOXK2 acts as a chromatin targeting factor for the BAP1-containing histone deubiquitinase complex. Nucleic Acids Res 42(10):6232–6242. doi:10.1093/nar/gku274
16. Kim J, Cantor AB, Orkin SH, Wang J (2009) Use of in vivo biotinylation to study protein-protein and protein-DNA interactions in mouse embryonic stem cells. Nat Protoc 4(4):506–517. doi:10.1038/nprot.2009.23
17. Kolodziej KE, Pourfarzad F, de Boer E, Krpic S, Grosveld F, Strouboulis J (2009) Optimal use of tandem biotin and V5 tags in ChIP assays. BMC Mol Biol 10:6. doi:10.1186/1471-2199-10-6
18. Byrum SD, Raman A, Taverna SD, Tackett AJ (2012) ChAP-MS: a method for identification of proteins and histone posttranslational modifications at a single genomic locus. Cell Rep 2(1):198–205. doi:10.1016/j.celrep.2012.06.019
19. Byrum SD, Taverna SD, Tackett AJ (2015) Purification of specific chromatin loci for proteomic analysis. Methods Mol Biol 1228:83–92. doi:10.1007/978-1-4939-1680-1_8
20. Byrum SD, Taverna SD, Tackett AJ (2013) Purification of a specific native genomic locus for proteomic analysis. Nucleic Acids Res 41(20):e195. doi:10.1093/nar/gkt822
21. Waldrip ZJ, Byrum SD, Storey AJ, Gao J, Byrd AK, Mackintosh SG, Wahls WP, Taverna SD, Raney KD, Tackett AJ (2014) A CRISPR-based approach for proteomic analysis of a single genomic locus. Epigenetics 9(9):1207–1211. doi:10.4161/epi.29919
22. Dejardin J, Kingston RE (2009) Purification of proteins associated with specific genomic Loci. Cell 136(1):175–186. doi:10.1016/j.cell.2008.11.045
23. Antao JM, Mason JM, Dejardin J, Kingston RE (2012) Protein landscape at Drosophila melanogaster telomere-associated sequence repeats. Mol Cell Biol 32(12):2170–2182. doi:10.1128/MCB.00010-12
24. Foulds CE, Feng Q, Ding C, Bailey S, Hunsaker TL, Malovannaya A, Hamilton RA, Gates LA, Zhang Z, Li C, Chan D, Bajaj A, Callaway CG,

Edwards DP, Lonard DM, Tsai SY, Tsai MJ, Qin J, O'Malley BW (2013) Proteomic analysis of coregulators bound to ERalpha on DNA and nucleosomes reveals coregulator dynamics. Mol Cell 51(2):185–199. doi:10.1016/j.molcel.2013.06.007

25. Mittler G, Butter F, Mann M (2009) A SILAC-based DNA protein interaction screen that identifies candidate binding proteins to functional DNA elements. Genome Res 19(2):284–293. doi:10.1101/gr.081711.108

26. Sharma NL, Massie CE, Butter F, Mann M, Bon H, Ramos-Montoya A, Menon S, Stark R, Lamb AD, Scott HE, Warren AY, Neal DE, Mills IG (2014) The ETS family member GABPalpha modulates androgen receptor signalling and mediates an aggressive phenotype in prostate cancer. Nucleic Acids Res 42(10):6256–6269. doi:10.1093/nar/gku281

27. Hata A, Seoane J, Lagna G, Montalvo E, Hemmati-Brivanlou A, Massague J (2000) OAZ uses distinct DNA- and protein-binding zinc fingers in separate BMP-Smad and Olf signaling pathways. Cell 100(2):229–240

28. Massie CE, Adryan B, Barbosa-Morais NL, Lynch AG, Tran MG, Neal DE, Mills IG (2007) New androgen receptor genomic targets show an interaction with the ETS1 transcription factor. EMBO Rep 8(9):871–878. doi:10.1038/sj.embor.7401046

29. Ross-Innes CS, Stark R, Holmes KA, Schmidt D, Spyrou C, Russell R, Massie CE, Vowler SL, Eldridge M, Carroll JS (2010) Cooperative interaction between retinoic acid receptor-alpha and estrogen receptor in breast cancer. Genes Dev 24(2):171–182. doi:10.1101/gad.552910

30. Sadar MD (1999) Androgen-independent induction of prostate-specific antigen gene expression via cross-talk between the androgen receptor and protein kinase A signal transduction pathways. J Biol Chem 274(12):7777–7783

31. Sharma NL, Massie CE, Butter F, Mann M, Bon H, Ramos-Montoya A, Menon S, Stark R, Lamb AD, Scott HE, Warren AY, Neal DE, Mills IG (2014) The ETS family member GABPα modulates androgen receptor signalling and mediates an aggressive phenotype in prostate cancer. Nucleic Acids Res 42(10):6256–6269

Chapter 10

Measuring the Expression of microRNAs Regulated by Androgens

Mauro Scaravilli, Kati Kivinummi, Tapio Visakorpi, and Leena Latonen

Abstract

The discovery of microRNAs (miRNAs) provided yet another mechanism of gene expression regulation. miRNAs have recently been also implicated in many diseases, including prostate cancer (PC). As PC is a highly androgen-dependent disease, extensive effort has been invested to identify the miRNAs that are androgen regulated. However, relatively few of them have been shown to be directly androgen regulated in PC. In this chapter we introduce the commonly used techniques to study the androgen regulation of miRNAs. The most cost-effective tool to profile global miRNA expression is microarray-based hybridization, whereas real-time quantitative reverse transcription PCR (qRT-PCR) is commonly used for the study of individual miRNAs.

Key words microRNA, Prostate cancer, Androgens, Androgen receptor, Gene expression, Microarray, qRT-PCR

1 Introduction

miRNAs have been found to be differentially expressed in many human malignancies [1], and it has been shown that they can exert tumor-suppressive and oncogenic functions [2, 3]. Several miRNAs play an important role in the development of prostate cancer (PC) [4]. For example miR-101 and miR-15a/16-1 [5, 6] have been shown to function as tumor suppressors, whereas miR-21, miR-221, and miR-32 function as androgen-regulated oncogenes, driving the castration-resistant PC phenotype [7–11].

Androgen receptor (AR) is known to target hundreds of genes, including several kallikreins, such as *KLK2* and *KLK3*, also known as PSA, and *TMPRSS2*. However, little is known about miRNAs that are directly targeted by AR. It seems that some of the androgen-regulated miRNAs are regulated through other mechanisms as well, thus making it difficult to assess the significance of the androgen dependence. For example, miR-21, which is known to be androgen regulated, is highly expressed in many AR-negative tissues as well.

Iain J. McEwan (ed.), *The Nuclear Receptor Superfamily: Methods and Protocols*, Methods in Molecular Biology, vol. 1443, DOI 10.1007/978-1-4939-3724-0_10,

Most of the studies on androgen regulation are based on the LNCaP cell line model. However, at least for miR-21, 29a, 29b, and 221 and for miR-17, 18b, 19b, 20a, 20b, 93, and 148a, androgen regulation has been observed in several AR-positive cell lines and xenografts as well [7, 12–14]. In addition, in at least one study [15] neoadjuvant treatment with goserelin and bicalutamide was used to identify androgen-regulated miRNAs in vivo in patients who had undergone radical prostatectomy. The published literature on androgen regulation of miRNAs often shows some discrepancies, which may be due to the use of different ligands, induction times, and cell lines. Relatively long androgen exposure time is needed to detect the induction of miRNA expression and this may be due to either slow biogenesis from the transcribed pri-miRNAs to pre- and mature forms by drosha and dicer enzymes (reviewed by Bartel [16]), or indirect androgen regulation mechanism. Yet another source of variability in the data is introduced by the method used to detect the gene expression. Some of the tools measure the mature form of the miRNAs specifically, whereas others can detect pre-miRNAs as well. miRNAs are also known to be polymorphic, and some of the detection tools, such as microarrays by Agilent Technologies, are sensitive to 3′ end length polymorphisms. In addition, different methods of normalization in microarray analyses and the use of different reference genes in qRT-PCR assays can easily result in discrepancies in the outcome.

We have recently identified miR-32 (officially hsa-miR-32-5p) as an androgen-regulated miRNA, which is overexpressed in castration-resistant prostate cancer [11]. The miR-32 expression data were obtained by both qRT-PCR and microarray in prostate cancer cell lines as shown in Fig. 1a. The direct androgen regulation was demonstrated with various means. Expression of miR-32 was shown to be induced by androgens (Fig. 1b), and the AR-binding site (ARBS) was first identified with chromatin immunoprecipitation-sequencing (ChIP-seq) [17], followed by validation with ChIP-PCR (Fig. 1c).

The introduction of hybridization techniques to detect the expression of mature miRNAs has expanded the number of studies on miRNAs. The microarrays are still the most cost-effective method to study large-scale miRNA expression, as reviewed by Li and Ruan [18]. And the qRT-PCR technique is still most commonly used for studying the expression of individual miRNAs. Other techniques worth mentioning are in situ hybridization and Northern blot, the latter providing the only reliable tool to validate the actual expression of mature miRNAs in some applications [19]. The new next-generation sequencing-based techniques provide yet another set of tools to detect gene expression as well as transcription factor binding and chromatin structure. However, these approaches lie out with the present discussion.

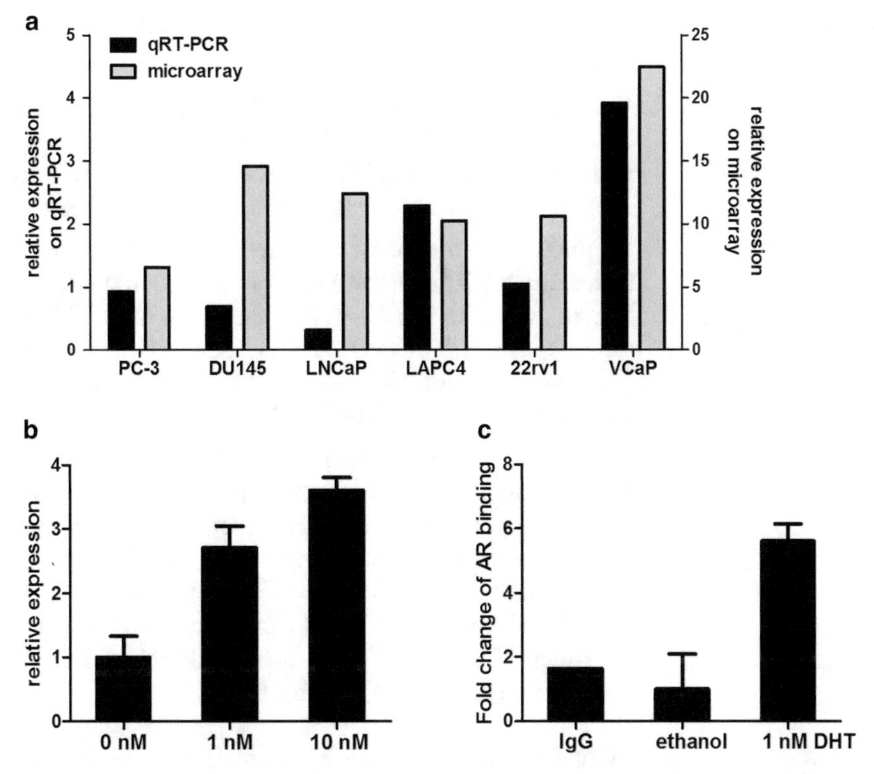

Fig. 1 miR-32 is an androgen regulated miRNA. (**a**) Comparison of qRT-PCR (TaqMan) and microarray (Agilent Technologies) analysis in measuring the expression levels of miR-32 in different prostate cancer cell lines. (**b**) Induction of miR-32 expression in LNCaP cells upon stimulation with 0, 1, and 10 nM of dihydrotestosterone (DHT) for 6 h. The expression of miR-32 was measured with qRT-PCR (TaqMan). The miR-32 values were normalized against RNU44 expression. (**c**) AR binds to an ARBS near miR-32 genomic location. LNCaP-ARhi cells stably expressing high levels of AR [21] were stimulated with 1 nM DHT for 2 h. ChIP-qPCR was performed [11]

In this chapter, we describe in detail how LNCaP prostate cancer cells are stimulated by androgens and used for measuring androgen-responsive miRNA expression by microarray (Agilent platform) and qRT-PCR (TaqMan assay).

2 Materials

Proper cell culturing techniques have the primary importance for the study of androgen regulation. For example, the use of different forms of androgens, as well as the exposure time and cell lines, affects the results. Use only sterile supplies, media, and reagents and work always under a cell culture hood when handling cells.

Prepare all solutions using ultrapure, sterile water and cell culture-grade reagents. Prepare and store all reagents at room temperature (unless indicated otherwise). Diligently follow all waste disposal regulations when disposing of waste materials. In RNA work use only RNase-free water and aseptic gloves. Avoid all possible RNase contamination and work promptly.

2.1 Cell Culturing Reagents

1. Use ATCC LNCaP PC cell line (Clone FGC, ATCC® CRL-1740™) (*see* **Note 1**).
2. Cell culturing medium for LNCaP cells: ATCC-formulated RPMI-1640 Medium (Catalog No. 30-2001), supplied with fetal bovine serum (FBS, Gibco®, Life Technologies™, Cat. No: 10270-106 or equivalent) to a final concentration of 10%.
3. Stripped medium for LNCaP cells to reduce androgens prior to treatment: RPMI-1640 without phenol red (Biowhittaker®, Lonza, Cat.No: BE12-918F) supplied with 10% charcoal dextran-stripped (CDS) FBS (Gibco®, Life Technologies™, Cat. No: 12676-029) and 2 mM l-glutamine (Gibco®, Life Technologies™, Cat. No: 25030-081).
4. Androgen treatment medium for LNCaP cells: RPMI-1640 without phenol red (Biowhittaker®, Lonza, Cat.No: BE12-918F) supplied with 10% charcoal dextran-stripped (CDS) FBS (Gibco®, Life Technologies™, Cat. No: 12676-029), 2 mM l-glutamine (Gibco®, Life Technologies™, Cat. No: 25030-081), and 1–100 nM dihydrotestosterone or synthetic androgen (DHT or R1881, Sigma-Aldrich, Cat.No: 730637 and R908, respectively) (*see* **Note 2**).

2.2 Reagents for Total-RNA Isolation

1. Trizol®-Reagent (Ambion, Life Technologies, Cat. No: 15596-018).
2. Chloroform.
3. Isopropyl alcohol.
4. 75% Ethanol (in DEPC-treated water).
5. RNase-free water.
6. Centrifuge and rotor capable of reaching up to 12,000 × *g*.
7. Polypropylene microcentrifuge tubes.

2.3 qRT-PCR Reagents

The most important aspect when studying small-RNA expression is to obtain reliable and specific detection of the RNA of interest. TaqMan® MicroRNA Assay (Thermo Fisher Scientific, Waltham, MA) allows the detection and quantification of specific miRNAs in 1–10 ng of total RNA and can distinguish the mature form from its precursor. The assay consists of preformulated primer and probe sets designed to detect and quantify mature miRNAs. The primers are available for the majority of the miRNAs included in the miRBase database, making the technology ideal for both high-throughput expression studies and validation of expression data acquired from microarray or sequencing platforms. The materials required for miRNA qRT-PCR using TaqMan® MicroRNA Assay are the following:

1. TaqMan® MicroRNA Reverse Transcription Kit (Applied Biosystems® 4366596).
2. TaqMan® Universal PCR Master Mix (Applied Biosystems® 4324018).

3. TaqMan® MicroRNA assay(s) (specific for the miRNAs of interest).
4. qRT-PCR thermal cycler.
5. qRT-PCR data analysis software.

2.4 Microarray Reagents

miRNA microarrays provide a cost-effective and high-throughput method for studying miRNA expression in cell lines, allowing the detection of all the known mature miRNAs as included in the database miRBase (http://mirbase.org/). Agilent Technologies (Santa Clara, CA) provides glass slides, each of them containing eight arrays of 60,000 probes per array and hybridization kit for the preparation of the sample to be used in the experiment. The reagents and materials required for Agilent miRNA microarray are the following:

1. miRNA Complete Labeling and Hybridization Kit (Agilent 5190-0456).
2. MicroRNA Spike-In Kit (Agilent 5190-1934, optional).
3. Gene Expression Wash Buffer Kit (Agilent 5188-5327).
4. Human miRNA Microarray slide, Release 19.0, 8x60K (Agilent G4872A-046064).
5. Microarray Scanner (Agilent G4900DA, G2565CA or G2565BA).
6. Agilent G4450AA Feature Extraction software 9.5 or later when Spike-In is not used or Feature Extraction 10.7.3 when Spike-In is included in the experiment.
7. Agilent Scan Control software, version A. 7.0 or later.
8. Hybridization Chamber (Agilent G2534A).
9. Hybridization Chamber gasket slides 8 microarrays/slide (Agilent G2534-60014).
10. Hybridization oven with 20 rpm capability and temperature set at 55 °C (Agilent G2545A).
11. Nuclease-free 1.5 mL microcentrifuge tubes.
12. Magnetic stir bar (×2) and magnetic stir plate.
13. Slide-staining dish, with slide rack (×3).
14. Circulating water baths or heat blocks set to 16 °C, 37 °C, and 100 °C.
15. Vacuum concentrator with heater.

3 Methods

Note that cell culture conditions vary for each cell type, and this protocol has been adjusted especially for LNCaP cell line. Maintain cells in a cell incubator (37 °C, 5% CO_2). Grow LNCaP cells in 60 or 100 mm polystyrene tissue culture dishes depending on the number of cells (the amount of RNA) required for the following methods.

3.1 Hormonal Treatment of Androgen-Sensitive LNCaP Cells

1. Divide LNCaP cells in 1:3 ratio after reaching 80% confluence (appr. 5×10^4 cells cm^2) directly into stripped growth medium onto two equal dishes (*see* **Notes 1**, **3**, and **4**).
2. Let the cells grow undisturbed in a cell incubator for 3–4 days.
3. Check the viability of the cells.
4. Carefully remove the strip medium and replace with androgen treatment and control (vehicle) medium (*see* **Note 2**).
5. Let both dishes be undisturbed for 24 h.
6. Collect total RNA from both androgen treatment and control dish as explained in next paragraph.

3.2 Isolation of Total RNA

1. Remove growth medium from culture dish.
2. Add 6 mL TRIzol® Reagent directly onto the cells in the 100 mm culture dish (*see* **Note 4**).
3. Work under hood and lyse the cells directly in the culture dish by pipetting up and down several times (*see* **Note 5**).
4. Collect and incubate the homogenized sample in an RNase-free polypropylene microcentrifuge tube for 5 min at room temperature (*see* **Note 6**).
5. Add 1.2 mL of chloroform.
6. Shake the tube vigorously by hand for 15 s.
7. Incubate for 2–3 min at room temperature.
8. Centrifuge the sample at $12,000 \times g$ for 15 min at 4 °C. After centrifugation the mixture will separate into a lower red phenol-chloroform phase, an interphase, and a colorless upper aqueous phase (~50% of the total volume and includes the total RNA).
9. Remove the aqueous phase of the sample by angling the tube and pipetting the solution out (*see* **Note** 7).
10. Place the aqueous phase into a new RNase-free tube.
11. Add 3 mL of 100% isopropanol to the aqueous phase.
12. Incubate at room temperature for 10 min.
13. Centrifuge at $12,000 \times g$ for 10 min at 4 °C.
14. Remove carefully the supernatant from the tube.
15. Wash the pellet, with 6 mL of 75% ethanol (*see* **Note 8**).
16. Vortex the sample briefly, and then centrifuge the tube at $7500 \times g$ for 5 min at 4 °C.
17. Carefully discard all the supernatant with a fine tip pipet taking care of preserving the integrity of the RNA pellet (*see* **Note 9**).
18. Air-dry the pellet for 5–10 min, at room temperature or 37 °C, or as long as the pellet starts to turn slightly transparent (*see* **Note 10**).

19. Resuspend the RNA pellet in 50–100 μL RNase-free water by pipetting the solution up and down several times and measure the concentration and 260/280 (*see* **Note 11**) ratio with, e.g., Nanodrop 2000 spectrophotometer. Check the integrity of the RNA by agarose gel electrophoresis.
20. Check the success of the androgen stimulation if needed by measuring expression of the androgen-regulated *KLK3* gene (*PSA*) of vehicle-treated (0 nM DHT) and DHT-treated RNA samples (*see* **Note 12**).

3.3 Microarray

In order to achieve a successful microarray hybridization it is important to assess the integrity of the RNA samples, either by agarose gel electrophoresis or by using an automated quantitation and quality control instrument such as Agilent Bioanalyzer 2100. The protocol for Agilent miRNA microarray supports only total RNA as starting material; therefore it is not advised to use purified miRNAs or small RNA fractions.

The first step of the protocol consists in labeling the total RNA and optionally the spike-in solutions, which serve as process controls to help distinguish biological data from experimental errors. Two spike-in solutions are available, called labeling spike-in and hybridization spike-in, and they are used as controls for the labeling and hybridization steps, respectively. The spike-in solutions are prepared as follows:

1. Dilute the spike-in stock solution 1:100 by adding 2 μL of stock solution into 198 μL of dilution buffer provided with the spike-in kit to obtain the first dilution (*see* **Note 13**).
2. Dilute the first dilution 1:100 to obtain the second dilution as described above.
3. Dilute the second dilution 1:100 to obtain the third dilution.

The labeling of the RNA is performed according to the following steps:

4. Dilute the total RNA to obtain a final concentration of 50 ng/μL in nuclease-free water. Add 2 μL of RNA (100 ng) to a 1.5 mL polypropylene tube.
5. Prepare the calf intestinal alkaline phosphatase (CIP) master mix by mixing 0.4 μL of 10× calf intestinal phosphatase buffer, 1.1 μL of labeling spike-in solution (third dilution, **step 3** above), and 0.5 μL of calf intestinal phosphatase per reaction (it is advisable to add 10–20% extra volume to compensate pipetting losses).
6. Add 2 μL of CIP master mix to each sample to obtain a total reaction volume of 4 μL, gently mixing by pipetting.
7. Incubate the samples at 37 °C in a circulating water bath or heat block for 30 min. This step will dephosphorylate the 3′-end of the original RNA, to allow for subsequent labeling.

8. Denature the samples (this allows the breaking of any secondary structure in the starting material) by adding 2.8 μL of 100% DMSO to each sample and then incubating at 100 °C in a circulating water bath or heat block for 5–10 min. Immediately transfer the samples on ice-water bath after denaturation to ensure that the RNA remains properly denatured.
9. Prepare the ligation master mix (this step will add a labeled cyanine3-pCp to the 3′-end of the dephosphorylated RNA) by mixing 1.0 μL of 10× T4 RNA ligase buffer, pre-warmed at 37 °C and cooled to room temperature, 3.0 μL of cyanine3-pCp and 0.5 μL of T4 RNA ligase per reaction. Add 4.5 μL of ligation master mix to each sample, mix gently by pipetting, and incubate at 16 °C for 2 h in a circulating water bath.

   After ligation, the samples can be optionally purified using MicroBioSpin 6 Columns (Bio-Rad 732-6221).

   After ligation or purification, the labeled RNA must be completely dried, using a vacuum concentrator for up to 3 h at 45–55 °C.

   The next step consists in hybridizing the labeled RNA to the microarray slide. The hybridization reaction is prepared as follows:
10. Resuspend the dried samples into 17 μL of nuclease-free water (18 μL if spike-in solution is not used).
11. Prepare the hybridization mix by adding 1.0 μL of hybridization spike-in solution (third dilution), 4.5 μL of 10× GE blocking agent, and 22.5 μL of 2× Hi-RPM hybridization buffer to each sample for a total volume of 45 μL.
12. Incubate the hybridization mix at 100 °C water bath or heat block for 5 min, immediately followed by 5 min on ice-water bath.
13. Prepare the hybridization assembly chamber and dispense the volume of sample in the gasket well of the gasket slide. It is advised to avoid introducing air bubbles at this step.
14. Place the microarray slide with the active side of the slide (where the probes are bound) facing the gasket slide.
15. Close the hybridization assembly and verify the absence of air bubbles by gently turning the assembly and observing the volume inside the gasket wells.
16. Hybridize the slide(s) in the hybridization oven at 55 °C for 20 h, at a rotational speed of 20 rpm.

   The next step consists in washing the slides in gene expression wash buffers 1 and 2 (*see* **Note 14**). The wash buffer 2 must be pre-warmed overnight in a slide-staining dish to 37 °C before use. The wash protocol consists of three steps:

17. The microarray slide is detached from the gasket slide in wash buffer 1.
18. The microarray slide is washed for 1 min at room temperature in wash buffer 1 in a slide-staining dish on a magnetic stirrer with continuous agitation.
19. The slide is subsequently washed in wash buffer 2 at 37 °C for 1 min as described above.

    To minimize the impact of environmental oxidants it is advised to scan the slide immediately after washing. The scanner is connected to a computer and controlled via Agilent Scan Control software. The scanner generates .tiff image files for each scanned slide and the position and identity of each spot on the image is then quantified and qualified via Agilent feature extraction software which converts the intensity of the signal in each spot into a numerical value, normalized against background signal, and organized in a .txt file. The .txt file can be subsequently analyzed to obtain expression values for each miRNA in the array slide and samples can be compared to experimental controls to obtain biologically significant data.

3.4 qRT-PCR

For optimal qRT-PCR performance, prepare the reaction in a dedicated area to avoid contamination from artificial templates and keep all the reagents on ice (*see* **Note 15**). Thaw the reagents on ice and vortex and centrifuge them briefly to properly resuspend them. Calculate the number of reactions needed for the amount of samples to be used.

The first step is the reverse transcription of the total RNA. The reverse transcription enzyme (reverse transcriptase) will convert the miRNA(s) of interest into cDNA, using specifically designed RT-primer(s).

1. Prepare the RT master mix as shown in the following list for a total reaction volume of 7 μL per sample (consider adding 10–20% extra volume to compensate pipetting losses):
   (a) 0.15 μL 100 mM dNTPs
   (b) 1.00 μL MultiScribe™ Reverse Transcriptase, 50 U/μL
   (c) 1.50 μL 10× Reverse transcription buffer
   (d) 0.19 μL RNase inhibitor, 20 U/μL
   (e) Nuclease-free water 4.16 μL
2. Add 1–10 ng of total RNA in a total volume of 5 and 3 μL of reverse transcription miRNA-specific primer per reaction, per miRNA, for a total final volume of 15 μL per reaction. Each reaction can be prepared in a 0.5 mL polypropylene tube of well of a 96-well plate.
3. Seal the tube(s) or well plate(s) and mix thoroughly by inversion of the solution, followed by a brief centrifugation to bring

the solution to the bottom of the tube or well. Incubate the tube(s) or well plate(s) on ice for 5 min and subsequently transfer to the thermal cycler, using the following parameters to program the reaction:

(a) 16 °C, 30 min

(b) 42 °C, 30 min

(c) 85 °C, 5 min

At this point, the reaction can be stored in the freezer (−15 to −25 °C) until the next step.

4. The second step is the actual PCR amplification of the cDNA prepared in the first step. It is recommended to perform each reaction in triplicate to ensure optimal reliability and to include non-template controls (NTC) to evaluate background signal and nonspecific amplification.

   Prepare the qRT-PCR master mix as in the following list for a total reaction volume of 18.67 μL per individual sample (add 10–20% extra volume to compensate for pipetting losses):

   (a) 1.00 μL TaqMan® microRNA Assay (20×)

   (b) 10.00 μL TaqMan® Universal PCR Master Mix II (2×)

   (c) 7.67 μL Nuclease-free water

5. Add 1.33 μL of reverse transcription product to each sample, for a total final volume of 20 μL. The reverse transcription product must be diluted 1:15 before qRT-PCR as to avoid concentrated reverse transcription by-products to interfere with the PCR amplification. Prepare each reaction in a 0.5 mL tube or well of a 96-well reaction plate. Seal the tube(s) or plate(s) and mix by inversion, followed by brief centrifugation. Load the tube(s) or plate(s) in the thermal cycler, using the following parameters for the reaction program:

   (a) 50 °C 2 min

   (b) 95 °C 10 min

   (c) 95 °C 15 s

   (d) 60 °C 60 s

   (e) Repeat **steps 3** and **4** for 40 cycles

   When quantifying miRNA expression levels, variation in the amount of starting material, sample preparation, and RNA extraction, as well as in reverse transcription efficiency, can introduce errors. Therefore, it is highly recommended to normalize the raw expression values to endogenous control genes to correct for potential biases. The endogenous control needs to be accurately validated and its expression needs to be relatively constant and abundant in the particular sample set used in the experiment (*see* **Note 16**).

4 Notes

1. LNCaP cell line is heterogeneous, and may thus lose its AR expression when cultured for extended periods of time in vitro. Use fresh ATCC clone or check the AR status with conventional qRT-PCR and/or western blot.
2. EtOH/DMSO concentration (solvent) of the ready medium should be less than 0.001%.
3. Start the procedure soon after reaching 80% cell confluence to avoid excessive clumping of the cells.
4. This protocol has been adjusted for 100 mm dish. If using smaller or larger plates adjust the volumes of reagents needed, or follow the manufacturer's guidelines.
5. Pipet the cells up and down as long as the Trizol® reagent is clear, free from fibrous-like filaments, to ensure proper lysis of the cells.
6. After this step, the homogenized sample can be stored at −20 °C for up to 1 week before proceeding. Ensure proper thawing and incubation at RT when continuing later with the isolation protocol.
7. Avoid collecting any of the interphase or organic layer into the pipette when removing the aqueous phase.
8. The RNA can be stored in 75% ethanol for at least 1 year at −20 °C.
9. Repeat the centrifugation or increase the centrifugation speed if needed to avoid detaching of the RNA pellet.
10. Do not allow the RNA to dry completely or use vacuum centrifuge, because fully dried RNA can lose its solubility.
11. In order to ensure accurate results, a ratio of ~2.0 is generally required to guarantee good purity of the RNA. If the ratio is significantly lower, it may indicate the presence of protein, phenol, or other contaminants that negatively affect the efficiency of microarray and qRT-PCR.
12. If necessary, before proceeding to MicroArray ensure the success of the androgen stimulation by assaying the well-known, strongly androgen-regulated *KLK3* (PSA) transcript, using conventional qRT-PCR; *see* Waltering et al. [21].
13. The first dilution of the spike-in can be stored at −80 °C for future use, although it is advisable to limit the freeze/thaw cycles to a maximum of two.
14. Add 2 mL of 10% Triton X-102, provided with the wash buffers, to both wash buffers 1 and 2 before use.

15. Particular care must be exercised as to avoid excessive exposure of TaqMan® microRNA Assay to light, as this might affect the fluorescent probe.
16. An example of reference gene validation is shown by Scaravilli et al. [20] in a prostate cancer clinical sample dataset. The authors compared qRT-PCR expression data of a subset of selected miRNAs with previously generated microarray and small RNA deep-sequencing data. The qRT-PCR results were normalized with four commonly used reference genes (RNU6B, RNU44, RNU24, and RNU48). The RNU6B-normalized expression results were the most consistent with microarray hybridization and deep-sequencing data. Moreover, the authors analyzed the individual expression of RNU44, RNU24, and RNU48 in the same set of samples, using RNU6B as a reference gene, confirming significant deregulation of all three genes in cancer samples, compared with the normal controls.

References

1. Rosenfeld N, Aharonov R, Meiri E et al (2008) MicroRNAs accurately identify cancer tissue origin. Nat Biotechnol 26:462–469
2. Kumar MS, Lu J, Mercer KL, Golub TR, Jacks T (2007) Impaired microRNA processing enhances cellular transformation and tumorigenesis. Nat Genet 39:673–677
3. Lotterman CD, Kent OA, Mendell JT (2008) Functional integration of microRNAs into oncogenic and tumor suppressor pathways. Cell Cycle 7:2493–2499
4. Catto JW, Alcaraz A, Bjartell AS et al (2011) MicroRNA in prostate, bladder, and kidney cancer: a systematic review. Eur Urol 59:671–681
5. Varambally S, Cao Q, Mani RS, Chinnaiyan AM et al (2008) Genomic loss of microRNA-101 leads to overexpression of histone methyltransferase EZH2 in cancer. Science 322:1695–1699
6. Bonci D, Coppola V, Musumeci M et al (2008) The miR-15a-miR-16-1 cluster controls prostate cancer by targeting multiple oncogenic activities. Nat Med 14:1271–1277
7. Ribas J, Ni X, Haffner M et al (2009) miR-21: an androgen receptor-regulated microRNA that promotes hormone-dependent and hormone-independent prostate cancer growth. Cancer Res 69:7165–7169
8. Ribas J, Lupold SE (2010) The transcriptional regulation of miR-21, its multiple transcripts, and their implication in prostate cancer. Cell Cycle 9:923–929
9. Sun T, Wang X, He HH et al (2013) MiR-221 promotes the development of androgen independence in prostate cancer cells via downregulation of HECTD2 and RAB1A. Oncogene 33:2790–2800
10. Sun T, Wang Q, Balk S et al (2009) The role of microRNA-221 and microRNA-222 in androgen-independent prostate cancer cell lines. Cancer Res 69:3356–3363
11. Jalava SE, Urbanucci A, Latonen L et al (2012) Androgen-regulated miR-32 targets BTG2 and is overexpressed in castration-resistant prostate cancer. Oncogene 31:4460–4471
12. Shi XB, Xue L, Yang J et al (2007) An androgen-regulated miRNA suppresses Bak1 expression and induces androgen-independent growth of prostate cancer cells. Proc Natl Acad Sci U S A 104:19983–19988
13. Ambs S, Prueitt RL, Yi M et al (2008) Genomic profiling of microRNA and messenger RNA reveals deregulated microRNA expression in prostate cancer. Cancer Res 68:6162–6170
14. Waltering KK, Porkka KP, Jalava SE et al (2011) Androgen regulation of micro-RNAs in prostate cancer. Prostate 71:604–614
15. Lehmusvaara S, Erkkilä T, Urbanucci A et al (2013) Goserelin and bicalutamide treatments alter the expression of microRNAs in the prostate. Prostate 73:101–112
16. Bartel DP (2009) MicroRNAs: target recognition and regulatory functions. Cell 136: 215–233

17. Urbanucci A, Sahu B, Seppälä J et al (2012) Overexpression of androgen receptor enhances the binding of the receptor to the chromatin in prostate cancer. Oncogene 31:2153–2163
18. Li W, Ruan K (2009) MicroRNA detection by microarray. Anal Bioanal Chem 394:1117–1124
19. Porkka KP, Pfeiffer MJ, Waltering KK et al (2007) MicroRNA expression profiling in prostate cancer. Cancer Res 67:6130–6135
20. Scaravilli M, Porkka KP, Brofeldt A et al (2015) MiR-1247-5p is overexpressed in castration resistant prostate cancer and targets MYCBP2. Prostate 75(8):798–805
21. Waltering KK, Helenius MA, Sahu B et al (2009) Increased expression of androgen receptor sensitizes prostate cancer cells to low levels of androgens. Cancer Res 69: 8141–8149

Chapter 11

Methods for Identifying and Quantifying mRNA Expression of Androgen Receptor Splicing Variants in Prostate Cancer

Yingming Li and Scott M. Dehm

Abstract

Constitutively active androgen receptor (AR) variants (AR-Vs) lacking the AR ligand-binding domain have been identified as drivers of prostate cancer resistance to AR-targeted therapies. A definitive understanding of the role and origin of AR-Vs in the natural history of prostate cancer progression requires cataloging the entire spectrum of AR-Vs expressed in prostate cancer, as well as accurate determination of their expression levels relative to full-length AR in clinical tissues and models of progression. Exon constituency differences at the 3′ terminus of mRNAs encoding AR-Vs compared with mRNAs encoding full-length AR can be exploited for discovery and quantification-based experiments. Here, we provide methodological details for 3′ rapid amplification of cDNA ends (3′ RACE) and absolute quantitative RT-PCR, which are cost-effective approaches for identifying new AR-Vs and quantifying their absolute expression levels in conjunction with full-length AR in RNA samples derived from various sources.

Key words Prostate cancer, Androgen receptor, Castration-resistant, Alternative splicing, AR splice variant, Absolute quantification, RT-PCR

1 Introduction

AR-Vs are composed of the transcriptionally active AR NH_2-terminal domain (NTD) and DNA-binding domain (DBD), but lack the regulatory ligand-binding domain (LBD) of the full-length AR protein. Multiple AR-Vs have been identified that arise from differential mRNA splicing of discrete 3′ terminal exons encoding novel COOH terminal amino acids and translation stop codons [1]. AR-Vs are expressed at the mRNA and protein level in prostate cancer cell lines, animal models, and clinical tissues [2–7]. Mechanistically, expression of AR-Vs in prostate cancer is due to rearrangements in the AR gene [8–10] as well as alterations in AR splicing regulation [11, 12]. As a result of harboring the AR NTD/DBD core, diverse AR-Vs can function as constitutively active transcription factors that are impervious to the spectrum of

Iain J. McEwan (ed.), *The Nuclear Receptor Superfamily: Methods and Protocols*, Methods in Molecular Biology, vol. 1443, DOI 10.1007/978-1-4939-3724-0_11,

AR-targeted therapies that exert their action through the intact AR LBD [13–15].

Diverse methods have been employed for discovery of suspected AR-Vs in prostate cancer cell lines and tissues. For example, related techniques such as 3′ rapid amplification of cDNA ends (RACE) and ligation-mediated PCR followed by Sanger sequencing were used to clone AR mRNA variants consisting of contiguously spliced AR exons 1/2/2b, 1/2/3/2b (also referred to as AR-V4), and 1/2/3/3b (also referred to as AR 1/2/3/CE3, AR-V7, or AR3) from the 22Rv1 prostate cancer cell line (Table 1 and Fig. 1a) [2, 3]. Similarly, 3′ RACE followed by Sanger sequencing or next-generation 454 and SOLiD sequencing was used to identify additional AR-V species expressed in human VCaP prostate cancer cells as well as the murine Myc-CaP prostate cancer cell line [7]. Interestingly, it appears that evidence for the existence of several AR-Vs may have been readily apparent from public gene expression databases. For example, several discrete 3′ terminal exons found in AR-V mRNAs, including cryptic exons (CE)1, CE2, CE3, and CE4 (Table 1 and Fig. 1a), had been identified as expressed sequence tags (ESTs) in previous gene expression studies [5]. With this information, Hu and colleagues designed PCR primers to amplify AR cDNAs harboring these predicted sequences from oligo dT-primed mRNA, leading to the identification of seven discrete AR-Vs, which were termed

Table 1
AR-VS identified in prostate cancer

AR-V name	Alternative name	Exon composition	Methods used for discovery	References
AR-V1	AR4	1/2/3/CE1	Informatics and RT-PCR, 3′ RACE	[3, 5]
AR-V2		1/2/3/3/CE1	Informatics and RT-PCR	[3, 5]
AR-V3	AR1/2/2b	1/2/2b	3′ RACE, informatics, and RT-PCR	[2, 3, 5]
AR-V4	AR1/2/3/2b, AR5	1/2/3/2b		[2, 3, 5]
AR-V5		1/2/3/CE2	Informatics and RT-PCR	[3]
AR-V6		1/2/3/CE2′		
AR-V7	AR3	1/2/3/CE3	Informatics and RT-PCR, 3′ RACE	[3, 5]
AR-V8			3′ RACE	[7]
AR-V9		1/2/3/V9		[7, 17]
AR-V10				[7, 17]
AR-V11				
AR-V12	AR^{v567es}	1/2/3/4/8	RT-PCR	[6, 16]
AR-V13		1/2/3/4/5/6/9	SLASR, tiling microarray	[16]
AR-V14		1/2/3/4/5/6/7/9		

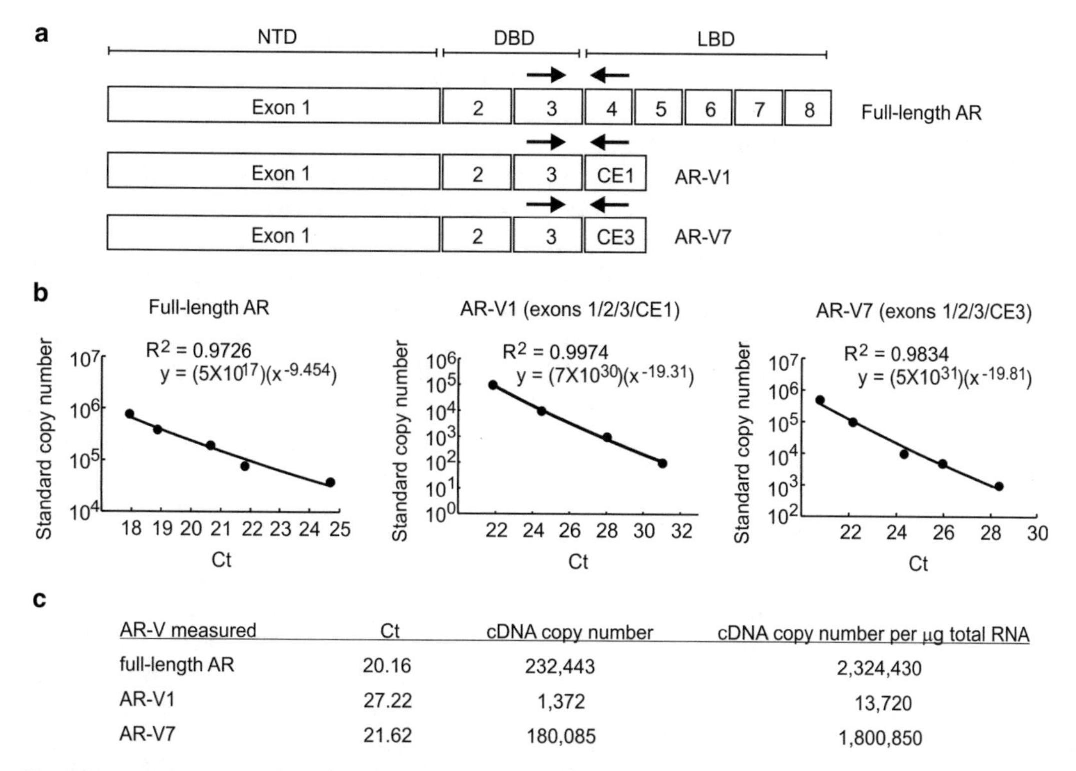

AR-V measured	Ct	cDNA copy number	cDNA copy number per μg total RNA
full-length AR	20.16	232,443	2,324,430
AR-V1	27.22	1,372	13,720
AR-V7	21.62	180,085	1,800,850

Fig. 1 Strategy for quantifying mRNA expression levels for full-length AR, AR-V1, and AR-V7. (**a**) Schematic of PCR pairs specific for discrete AR cDNA species. (**b**) Copy number vs. threshold cycle of amplification (Ct) standard curves constructed using PCR primer pairs in (a) and plasmid templates harboring the cDNA of interest. (**c**) Representative calculations for conversion of Ct values to cDNA copy numbers per μg of RNA for individual AR species expressed in the CWR-R1 cell line

AR-V1 through AR-V7 [5]. Using a related targeted PCR approach, Sun and colleagues discovered AR^{v567es} arising from direct splicing of AR exon 4 to AR exon 8. Finally, expression of a novel AR exon 9 and AR-Vs including AR-V13 and AR-V14 was identified using selective linear amplification of sense RNA (SLASR) and hybridization to microarrays containing 60-mer probes tiled across the human AR gene locus [16]. Table 1 provides a summary of AR-V species that have been identified to date, including alternative nomenclatures, exon composition, and methods used for initial discovery.

In addition to methods adapted for AR-V discovery purposes, various methods have been developed for quantification of AR-V mRNAs expressed in prostate cancer. Overall, quantitative RT-PCR (qRT-PCR) has been the mainstay technique used to measure levels of AR-V mRNAs in prostate cancer cell lines and tissues. Typical qRT-PCR approaches have involved determining the threshold cycles (Ct) of amplification for an AR mRNA species of interest and an internal control (such as GAPDH, 18S rRNA) in a series of RNA samples from prostate cancer cell lines or tissues [3, 5]. The $2^{-\Delta\Delta Ct}$ data transformation method can then be used to determine fold differences in expression levels of the mRNA target relative to the internal control across the samples studied [3, 5]. However, this strategy is not appropriate for determining expression levels of multiple AR mRNA targets relative to each other within a single sample. In this case, absolute quantification should be pursued, which involves the construction of Ct vs. copy number standard curves from plasmids harboring the appropriate AR PCR target sequence. Standard curves must be developed for each AR or AR-V mRNA target of interest, as this allows for extrapolation of cDNA copy number from the Ct value derived for each target in a given reverse-transcribed RNA sample [2, 8, 9]. These RT-PCR-based strategies have been important for addressing the clinical relevance of AR-Vs in prostate cancer progression and development of resistance to AR-targeted therapies. For example, AR-V1 and AR-V7 mRNA expression levels are higher in castration-resistant prostate cancer tissues than hormone-naïve prostate cancer tissues [5]. Additionally, survival of patients with castration-resistant prostate cancer following metastasis surgery was significantly shorter for those patients with the highest AR-V7 mRNA levels in their skeletal metastases [4]. While these data clearly illustrate an important role for certain AR-Vs in castration-resistant prostate cancer, the role of AR-Vs in untreated prostate cancer has been less clear. For example, in one study, RT-PCR-based assessment of AR-V mRNA levels in prostatectomy specimens demonstrated that high AR-V7 mRNA expression (but not AR-V1 mRNA expression) was associated with increased risk of disease recurrence [5]. However, a separate study employing a branched DNA assay for quantification of AR-V mRNAs found that neither AR-V1 nor AR-V7 mRNA levels were predictive of recurrence [17]. This indicates that the sensitivity and specificity of methods for AR-V mRNA quantification may have major impact on the ultimate conclusions that are derived from studies with clinical tissues.

Overall, deployment of these myriad methods has revealed that further discovery- and quantification-based studies are required, as is ongoing method development and optimization. To this end, this chapter outlines detailed protocols for (1) 3′ RACE methods for discovery of AR-Vs, and (2) absolute quantification RT-PCR methods for determining copy numbers of discrete AR and AR-V mRNA species in a single-RNA sample.

2 Materials

2.1 Total Cellular RNA Extraction

1. PBS: 3.2 mM Na_2HPO_4, 0.5 mM KH_2PO_4, 1.3 mM KCl, 135 mM NaCl, pH 7.4. Filter sterilize or autoclave and store at room temperature.
2. Solution D: 4 M Guanidinium thiocyanate, 25 mM sodium citrate, pH 7.0, 0.5%(w/v), *N*-laurosylsarcosine (sarkosyl),

0.1 M β-mercaptoethanol. Dissolve 47.26 g guanidinium thiocyanate in 70 mL of (DEPC)-treated H_2O (*see* **Note 1**), add 2.5 mL 1 M sodium citrate and 5 mL 10% sarcosyl, and add DEPC-treated H_2O to a final volume of 100 mL. Store at room temperature. Just prior to use, add 72 μL β-mercaptoethanol per 10 mL Solution D.

3. 2 M Sodium acetate, pH 4.0: Dissolve 16.42 g anhydrous sodium acetate in 35 mL H_2O. With stirring, add 35 mL glacial acetic acid and continue adjusting pH with glacial acetic acid to pH 4.0. Add H_2O to a final volume of 100 mL. Filter sterilize or autoclave and store at room temperature.
4. Phenol solution saturated with 0.1 M citrate buffer, pH 4.3.
5. Chloroform:isoamyl alcohol (49:1) (v/v): Mix 49 mL chloroform and 1 mL isoamyl alcohol. Store at 4 °C in a foil-wrapped bottle.
6. Ethanol, absolute (100%) or 95%.
7. RNA storage buffer: 0.1 mM EDTA, pH 8.0 in DEPC-treated water. Add 10 μL 0.5 M EDTA (pH 8.0) to 50 mL DEPC-treated H_2O, and store at 4 °C.
8. NanoDrop 200 spectrophotometer (Thermo Scientific or equivalent).

2.2 3′ RACE

1. 3′ RACE kit (Roche Applied Science or other equivalent).
2. AR gene-specific primers for 3′ RACE reaction, dissolved in H_2O at 12.5 μM final concentration. First-step PCR: AR Exon1 forward primer 5′-TTGAACTGCCGTCTACCCTGTC-3′, second-step nested PCR: AR Exon 1 forward primer 5′-ACAA CTTTCCACTGGCTCTGGC-3′.
3. Taq DNA polymerase (5 U/μL), 10× PCR reaction buffer (Qiagen or other equivalent).

2.3 Absolute Quantitative RT-PCR

1. cDNA synthesis kit: Transcriptor First Strand cDNA Synthesis Kit (Roche Applied Science or other equivalent).
2. Thin-walled 0.2 mL PCR tubes, certified RNase, DNase free.
3. PerfeCTa SYBR Green FastMix (Quanta Bioscience or equivalent).
4. 96-Well PCR plate and adhesive optical tape (Bio-Rad).
5. Forward and reverse primers specific for each AR mRNA species: Table 2 lists forward and reverse primer pairs that are used for quantifying levels of full-length AR and specific AR-Vs. All primers are dissolved in H_2O of 5 μM final concentration.
6. Thermocycler with fluorescence detector (Bio-Rad iCycler or equivalent).
7. Plasmids harboring full-length AR, AR1/2/2b, AR1/2/3/2b, AR 1/2/3/CE1, AR1/2/3CE2, and 1/2/3/CE3.

Table 2
Primer quantitative RT-PCR analysis of AR and AR-V expression

qPCR target	Primer sequence
AR-V1	FWD 5′-AAC AGA AGT ACC TGT GCG CC-3′
	RV 5′-TGA GAC TCC AAA CAC CCT CA-3′
AR-V5	FWD 5′-AAC AGA AGT ACC TGT GCG CC-3′
	RV 5′-TAT GAC ACT CTG CTG CCT GC-3′
AR-V7	FWD 5′-AAC AGA AGT ACC TGT GCG CC-3′
	RV 5′-TCA GGG TCT GGT CAT TTT GA-3′
AR-V4	FWD 5′-AAC AGA AGT ACC TGT GCG CC-3′
	RV 5′-TTC TGT CAG TCC CAT TGG TG-3′
AR-V3	FWD 5′-TGG ATG GAT AGC TAC TCC GG-3′
	RV5′-GTT CAT TCT GAA AAA TCC TTC AGC-3′
Full-length AR	FWD 5′-AAC AGA AGT ACC TGT GCG CC-3′
	RV 5′TTC AGA TTA CCA AGT TTC TTC AGC-3′

3 Methods

3.1 Total Cellular RNA Extraction

Total RNA can be prepared from cell lines and tissues using a simple and cost-effective acid-guanidinium phenol/chloroform extraction method [18] (*see* **Note 2**). The scale for a standard RNA extraction is a single 10 cm tissue culture plate. This can be scaled up/down as needed by adjusting reagent volumes by a factor equal to the ratio of the surface area of a 10 cm plate to the surface area of the tissue culture vessel being used. When extracting RNA from tissues, frozen tissue pieces should be pulverized under liquid nitrogen using a mortar and pestle.

1. Grow cells to 80% confluence on 10 cm tissue culture plates. Aspirate medium and wash cells once gently with 5 mL of ice-cold PBS. Aspirate the PBS.
2. Lyse washed cells by adding 500 μL of Solution D to the middle of the plate. Cells will lyse immediately when contacted by Solution D, making the solution quite viscous. Use a plastic scraper to distribute the viscous solution over the entire surface of the plate to lyse all cells. Transfer the lysate to a 1.5 mL microcentrifuge tube on ice. Pipet the lysate up and down ten times to shear genomic DNA and reduce viscosity.
3. Add 50 μL of 2 M sodium acetate, pH 4.0. Mix vigorously by vortex on high speed, for 10 s. Add 500 μL of the bottom, organic layer of the citrate-saturated phenol. Mix vigorously by

vortex on high speed, for 10 s. Add 100 μL of chloroform:isoamyl alcohol (49:1), and mix vigorously by vortex on high speed, for 10 s. Incubate on ice for 15 min, mixing vigorously by vortex for 10 s every 5 min. Separate aqueous and organic layers by centrifugation for 5 min at 14,000 × *g*, 4 °C.

4. Using a p200 pipet tip and a p200 pipettor, carefully transfer 200 μL of the top aqueous layer to a fresh tube, taking care to avoid the bottom organic layer as well as the white interface. Repeat this transfer for total volume extracted of 400 μL.
5. Precipitate RNA from the aqueous fraction by addition of 1 mL 95–100% ethanol. Mix well and incubate in a –20 °C freezer for at least 1 h. Pellet RNA by centrifugation for 15 min at 14,000 × *g*, 4 °C. Remove all ethanol, and partially air-dry the pellet for 5–10 min, taking care not to overdry.
6. Dissolve RNA in a suitable volume (usually 60 μL) of RNA storage buffer. Assess the concentration and purity of the RNA by UV spectrophotometry using a NanoDrop 200 spectrophotometer. A ratio of the absorbances measured at 260 and 280 nm that is 1.8 or greater can be considered pure. RNA meeting this quality threshold should be analyzed by electrophoresis in 1% agarose gels to ensure integrity of the 28S, 18S, and 5.8S ribosomal RNA bands.

3.2 3′ RACE

Total RNA isolated from prostate cancer cell lines or tissues can be subjected to 3′ RACE, with the goal of identifying 3′ terminal sequence of AR-V mRNAs. Full-length AR and AR-Vs all contain the AR NTD/DBD core encoded by AR exons 1, 2, and 3. Therefore an AR-anchored forward primer that binds within one of these AR exons is suitable for this approach. The sensitivity of 3′ RACE can be enhanced by using a nested PCR approach, wherein the first PCR reaction is used as input in a second PCR reaction with a nested AR-specific forward primer. The 3′ RACE PCR products can be subjected to next-generation sequencing, followed by mapping of reads to a reference human genome assembly to identify the genomic origin of the 3′ sequences [7]. A simpler and more widely available approach is to clone the 3′ RACE PCR products into plasmid vectors and conduct Sanger sequencing to identify the genomic origin of the 3′ sequences [2, 3].

3.2.1 First-Strand cDNA Synthesis

1. Add the following components from the 3′ RACE kit in a PCR tube on ice: 4 μL cDNA synthesis buffer, 2 μL deoxynucleotide mixture, 1 μL oligo-dT primer, 1 μg of total cellular RNA prepared in Subheading 3.1, 1 μL reverse transcriptase, and H_2O to total volume 20 μL. Calculate the volume of H_2O that will be required in the reaction prior to adding reaction components, and add this to the tube first. Next, add all reaction components except the reverse transcriptase enzyme, and finger-flick

gently to mix. Add the reverse transcriptase enzyme last, mix the reaction again by finger-flicking, and centrifuge briefly (2 s, 14,000 × *g*, room temperature).

2. Incubate in a thermocycler set for 60 min at 55 °C, 5 min at 85 °C, and a hold at 4 °C. The reaction can be left on hold at 4 °C overnight if desired. This reverse transcription reaction can be used immediately for PCR amplification, or stored at −20 °C.

3.2.2 First-Step PCR Amplification

1. Add the following components from the 3′ RACE kit in a PCR tube on ice: 1 μL cDNA from Subheading 3.2.1, 1 μL PCR anchor primer, 1 μL AR Exon1 forward primer (*see* Subheading 2.2), 1 μL deoxynucleotide mixture, 0.5 μL Taq DNA polymerase (5 U/μL), 5 μL 10× reaction buffer, and H_2O to a final reaction volume of 50 μL. Calculate the volume of H_2O that will be required in the reaction prior to adding reaction components, and add this to the tube first. Next, add all reaction components except the Taq DNA polymerase enzyme, and finger-flick gently to mix. Add the Taq DNA polymerase enzyme last, mix the reaction again by finger-flicking, and centrifuge briefly (2 s, 14,000 × *g*, 4 °C).
2. Place reaction tube in a thermocycler with the following step settings: step 1 (1 cycle): 95 °C for 2 min; step 2 (35 cycles): 95 °C for 30 s, 60 °C for 30 s, and 72 °C for 1 min; step 3: 72 °C for 1 min; step 4: hold at 4 °C. The reaction can be left on hold at 4 °C overnight if desired. This PCR product can be used immediately for nested PCR amplification or stored at −20 °C.

3.2.3 Nested PCR Amplification

There is a tendency for the standard 3′ RACE assay as outlined in Subheadings 3.2.1 and 3.2.2 to yield nonspecific PCR products. Because the goal of this approach is to identify mRNAs encoding AR-Vs, which may be expressed at levels significantly lower than mRNAs encoding full-length AR mRNA, a second nested PCR reaction can be beneficial. This second nested PCR reaction is performed with a nested AR forward primer, thus enriching for AR-derived mRNA species.

1. Dilute the first-step PCR amplification reaction 1:1000 with H_2O. This diluted PCR product is used as the template in the second nested PCR reaction.
2. Add the following components from the 3′ RACE kit in a PCR tube on ice: 1 μL diluted first-step PCR amplification product from Subheading 3.2.2, 1 μL PCR anchor primer, 1 μL nested AR Exon1 forward primer (*see* Subheading 2.2), 1 μL deoxynucleotide mixture, 0.5 μL Taq DNA polymerase (5 U/μL), 5 μL 10× reaction buffer, and H_2O to a final reaction

volume of 50 μL. Calculate the volume of H_2O that will be required in the reaction prior to adding reaction components, and add this to the tube first. Next, add all reaction components except the Taq DNA polymerase enzyme, and finger-flick gently to mix. Add the Taq DNA polymerase enzyme last, mix the reaction again by finger-flicking, and centrifuge briefly (2 s, 14,000 × *g*, 4 °C).

3. Place reaction tube in a thermocycler with the following step settings: step 1 (1 cycle): 95 °C for 2 min; step 2 (35 cycles): 95 °C for 30 s, 60 °C for 30 s, and 72 °C for 1 min; step 3: 72 °C for 1 min; step 4: hold at 4 °C. The reaction can be left on hold at 4 °C overnight if desired. This PCR product can be used immediately for cloning and sequencing or stored at −20 °C.
4. The final PCR product should contain a representative mixture of AR-derived mRNAs expressed in the original cell line or tissue. The predominant AR mRNA is expected to be full-length AR, with AR-Vs expressed at lower levels. Therefore, when considering downstream strategies for analyzing these products, it will be necessary to tailor the approach to the goal of the experiment. For example, if searching for a suspected AR-V mRNA that represents 10% of overall AR expression, it would be necessary to analyze ten individual colonies to identify just one clone harboring sequence from that suspected species.

3.3 Absolute Quantitative RT-PCR

In situations where the AR-V of interest is known, it is often important to quantify the expression levels of this AR-V relative to full-length AR in a particular cell line or tissue. Because different PCR primers are used to amplify different AR mRNA species, absolute quantitation is required for this approach.

3.3.1 Primer Design

PCR primer sets that discriminate specific known AR-Vs and full-length cDNAs can be designed using Primer3 software. The general strategy is to design a "universal" forward primer that binds within AR exon 2 or 3, and pair this forward primer with AR- and AR-V-specific reverse primers that bind within AR-V unique exons 2b, CE1, CE2, or CE3 (Fig. 1a). The sequences of primer pairs specific for discrete AR species that have been derived using this approach are outlined in Table 2. The working concentration for these primers is 5 μM in water, and they can be stored at −20 °C.

3.3.2 RT-PCR with Test RNA and AR Standards

For each individual AR mRNA species to be evaluated, it is necessary to set up a PCR reaction with cDNA prepared from the RNA sample of interest, as well as a set of serial dilutions of a plasmid or other cDNA sequence harboring the PCR amplicon for that individual AR mRNA species (Fig. 1b). For instance, if the goal of the experiment is to perform absolute quantification of full-length AR, AR-V1, and AR-R7 mRNA levels in the CWR-R1 prostate cancer cell line, PCR with primer pairs specific for these individual species will be performed on cDNA prepared from CWR-R1 cells as well as dilution series of plasmids encoding full-length AR, AR-V1, and AR-V7.

1. Perform reverse transcription reactions with a Transcriptor First Strand cDNA Synthesis Kit using 1 μg of total cellular RNA prepared in Subheading 3.1 as input. For the Transcriptor First Strand cDNA Synthesis Kit, the total reaction volume is 20 μL.
2. Prepare serial dilutions of plasmid DNA harboring the PCR amplicon of interest. For plasmid DNA at a concentration of 40 ng/μL, a typical dilution series is performed in 1:5 or 1:10 steps down to 10^{-12} of this original template, with all dilutions in H_2O (*see* **Note 3**).
3. Set up a single quantitative PCR reaction for each cDNA being evaluated as well as each standard in the plasmid dilution series. To one well of a 96-well PCR plate, add 10 μL 2× SYBR Green FastMix, 6 μL H_2O, 1 μL of forward primer, 1 μL of reverse primer, and input DNA (2 μL of a diluted plasmid standard or 2 μL of cDNA sample).
4. Seal the plate with adhesive optical tape, and centrifuge at 1600 × *g*, 4 °C for 5 min.
5. Place plates in a thermocycler with fluorescence detector with the following step settings: step 1 (1 cycle), 95 °C for 2 min; step 2 (40 cycles), 95 °C for 15 s and 60 °C for 15 s. Collect fluorescence intensity measurements during each of the 40 cycles, and determine Ct values for individual wells using the maximum curvature approach (*see* **Note 4**).
6. Construct Ct vs. copy number standard curves for each AR plasmid template used. To calculate copy number for each standard in a dilution series, the plasmid concentration and plasmid size (in base pairs) of each standard must be known. These data are used as input in Eq. 1:

$$\text{Copy number} = \frac{(\text{plasmid concentration} \quad V \quad 6.022 \quad 10^{23})}{(\text{plasmid length} \quad 650)} \qquad (1)$$

where plasmid concentration = concentration of standard (in g/μL), V = volume of plasmid template added to PCR reaction (2 μL as per this method), plasmid length = length of plasmid template added (in bp), and 650 = average molecular mass of a base pair (in g/mol).

Plot data as *XY* scatter charts in Microsoft Excel or similar graphing program, plotting Ct values on a logarithmic scale *X*-axis and copy number on the *Y*-axis. Generate a power trendline in the chart and display the *R*-squared value as well as the equation of this trendline. If the *R*-squared value is higher than 0.95, then this equation of the trendline can be used to calculate copy number from Ct data for the cDNA of interest (Fig. 1b).

7. Calculate cDNA copies generated per μg total RNA. Using the equation of the trendline determined from the standard curve, solve for copy number of a particular AR species in the cDNA sample of interest (*see* **Note 5**). For instance, if the copy numbers of AR-V7 cDNA are being determined, then the Ct determined for that cDNA sample with the AR-V7-specific primer pair is extrapolated against the trendline derived using AR-V7 plasmid dilutions. Because 2 μL of cDNA out of a 20 μL cDNA reaction was added to a well during setup of PCR reactions, the calculated copy number needs to be multiplied by a factor of 10 to derive the cDNA copy number generated per 1 μg RNA (Fig. 1c).

4 Notes

1. All water used in the experiment procedure should be of molecular-biology grade DEPC-treated or certified nuclease-free.
2. RNA prepared with alternative RNA extraction reagents such as Trizol (Life Technologies), or RNA extraction kits such as RNeasy (Qiagen), also performs very well in downstream assays.
3. It is often helpful to carry out an initial pilot experiment to determine the optimal range of plasmid dilutions that should be used to construct standard curves. The goal of this pilot experiment is to ensure that the Ct values for the AR species being measured fall within the range of the standards on the curve. This can be accomplished by constructing an initial broad standard curve from 1:10 dilutions, and determining where in this range the Ct value falls for the AR mRNA species of interest in the sample of interest. A second, narrower range of five standards can then be prepared as 1:5 or 1:2 dilutions, and used to construct the standard curves in subsequent experiments.
4. It is important to validate each PCR product by electrophoresis in 1% agarose gels to ensure that the Ct values determined for each well are the result of correct amplification of the desired single product. If multiple bands are visible, or bands of an incorrect size are observed, then Ct data for that well should not be used for subsequent steps.

5. Multiple technical and biological replicates are required to derive accurate measurements of expression of individual AR species in a cell line or tissue of interest. When the source of RNA is not limiting (for example, with cell lines), performing three biological replicate experiments, each in technical triplicate ($n = 9$ total), provides reliable measures of expression levels in that source as well as an accurate estimate of standard error in the measurement. When the source of RNA is limiting (as is often the case with tissues), performing three independent cDNA synthesis reactions from a single RNA sample, and then performing triplicate PCR reactions with these cDNA samples ($n = 9$ total), provides reliable measures of expression levels in that RNA sample as well as an accurate estimate of standard error in the measurement.

Acknowledgements

Grant Support: Studies in the Dehm Lab are supported by NCI Grant R01CA174777 (to S.M.D.) and an American Cancer Society Research Scholar Grant RSG-12-031-01-TBE (to S.M.D.).

References

1. Dehm SM, Tindall DJ (2011) Alternatively spliced androgen receptor variants. Endocr Relat Cancer 18:R183–R196
2. Dehm SM et al (2008) Splicing of a novel androgen receptor exon generates a constitutively active androgen receptor that mediates prostate cancer therapy resistance. Cancer Res 68:5469–5477
3. Guo Z et al (2009) A novel androgen receptor splice variant is up-regulated during prostate cancer progression and promotes androgen depletion-resistant growth. Cancer Res 69:2305–2313
4. Hornberg E et al (2011) Expression of androgen receptor splice variants in prostate cancer bone metastases is associated with castration-resistance and short survival. PLoS One 6:e19059
5. Hu R et al (2009) Ligand-independent androgen receptor variants derived from splicing of cryptic exons signify hormone-refractory prostate cancer. Cancer Res 69:16–22
6. Sun S et al (2010) Castration resistance in human prostate cancer is conferred by a frequently occurring androgen receptor splice variant. J Clin Invest 120:2715–2730
7. Watson PA et al (2010) Constitutively active androgen receptor splice variants expressed in castration-resistant prostate cancer require full-length androgen receptor. Proc Natl Acad Sci U S A 107:16759–16765
8. Li Y et al (2011) Intragenic rearrangement and altered RNA splicing of the androgen receptor in a cell-based model of prostate cancer progression. Cancer Res 71:2108–2117
9. Li Y et al (2012) AR intragenic deletions linked to androgen receptor splice variant expression and activity in models of prostate cancer progression. Oncogene 31:4759–4767
10. Nyquist MD et al (2013) TALEN-engineered AR gene rearrangements reveal endocrine uncoupling of androgen receptor in prostate cancer. Proc Natl Acad Sci U S A 110: 17492–17497
11. Liu LL et al (2013) Mechanisms of the androgen receptor splicing in prostate cancer cells. Oncogene 33(24):3140–3150
12. Yu Z et al (2014) Rapid induction of androgen receptor splice variants by androgen deprivation in prostate cancer. Clin Cancer Res 20(6):1590–1600
13. Chan SC et al (2012) Androgen receptor splice variants activate AR target genes and support aberrant prostate cancer cell growth independent of the canonical AR nuclear localization signal. J Biol Chem 287:19736–19749

14. Li Y et al (2013) Androgen receptor splice variants mediate enzalutamide resistance in castration-resistant prostate cancer cell lines. Cancer Res 73:483–489
15. Mostaghel EA et al (2011) Resistance to CYP17A1 inhibition with abiraterone in castration-resistant prostate cancer: induction of steroidogenesis and androgen receptor splice variants. Clin Cancer Res 17:5913–5925
16. Hu R et al (2011) A snapshot of the expression signature of androgen receptor splicing variants and their distinctive transcriptional activities. Prostate 71(15):1656–1667
17. Zhao H et al (2012) Transcript levels of androgen receptor variant AR-V1 or AR-V7 do not predict recurrence in patients with prostate cancer at indeterminate risk for progression. J Urol 188:2158–2164
18. Chomczynski P, Sacchi N (1987) Single-step method of RNA isolation by acid guanidinium thiocyanate- phenol-chloroform extraction. Anal Biochem 162:156–159

Part IV

Model Systems

Chapter 12

Harvesting Human Prostate Tissue Material and Culturing Primary Prostate Epithelial Cells

Fiona M. Frame, Davide Pellacani, Anne T. Collins, and Norman J. Maitland

Abstract

In order to fully explore the biology of a complex solid tumor such as prostate cancer, it is desirable to work with patient tissue. Only by working with cells from a tissue can we take into account patient variability and tumor heterogeneity. Cell lines have long been regarded as the workhorse of cancer research and it could be argued that they are of most use when considered within a panel of cell lines, thus taking into account specified mutations and variations in phenotype between different cell lines. However, often very different results are obtained when comparing cell lines to primary cells cultured from tissue. It stands to reason that cells cultured from patient tissue represents a close-to-patient model that should and does produce clinically relevant data. This chapter aims to illustrate the methods of processing, storing and culturing cells from prostate tissue, with a description of potential uses.

Key words Prostate, Tissue, Primary epithelial cells, Cancer, Benign prostatic hyperplasia

1 Introduction

Here we present methods to process prostate tissue from patient samples. We describe how to preserve tissue and how to obtain primary epithelial cells and stromal cells. We also describe how this material can be cultured and manipulated to recapitulate the tumor microenvironment in 2D and briefly in 3D. This method was adapted from [1, 2] and has been used in multiple studies from our laboratory [3–15]. In addition, several labs are using this or similar methods to grow primary prostate cells and work with prostate tissue [16–29].

Although difficult to obtain and maintain, primary cells offer several advantages over cell line models for the study of stem cell (SC) and cancer biology. Studies involving primary cells are often conducted on several samples obtained from independent patients, which recapitulates the natural interindividual heterogeneity. Thus, it is less likely for the results obtained to be a cell line-specific

Iain J. McEwan (ed.), *The Nuclear Receptor Superfamily: Methods and Protocols*, Methods in Molecular Biology, vol. 1443, DOI 10.1007/978-1-4939-3724-0_12, © Springer Science+Business Media New York 2016

phenomenon. In addition, there have been several recent reports that acknowledge and report misidentification and cross-contamination of cell lines, leading to discrepancies and non-reproducibility of results [30–32]. Indeed, to overcome this, publishing requirements are becoming more stringent, and cell lines have to be authenticated prior to publishing [33]. The use of primary cells derived from individual patients overcomes this problem.

Due to the limited growth potential of primary cells, they remain genetically and epigenetically close to the original tissue/tumor, with fewer chances for the acquisition of specific mutations and or cross contamination with cell lines or other samples [9]. However, there is no room for complacency and it goes without saying that primary cells should be cultured and stored very much separately from cell lines in a lab that has both. Cell lines have the advantage of being readily available and with the potential to grow unlimited numbers of cells for use in assays that require large amounts of cells (e.g., compound/shRNA library screenings, RNA sequencing, Chip-sequencing). There is now increasing interest and research effort into scaling down these kinds of assays such that they can be carried out on small populations of selected cells from primary cultures or directly from tissue.

Primary cells have already proven their worth in pushing treatments towards clinical trials. In one study to establish the specificity of a prostate-targeted oncolytic adenovirus a panel of primary cells, including prostate, were used for preclinical specificity testing [12]. As a result, this virus is now in a phase I trial [34]. In addition, studies have shown that when testing chemotherapeutic drugs, primary cells give very different results to cell lines and typically have up to a log higher IC50 values [35].

It could be thought that the limited growth potential of primary cells may be a disadvantage and that they may be less amenable to manipulation than cell lines. However, practical methods have progressed such that transfection reagents for introducing plasmids and siRNA are more efficient and technology such as TALEN and CRISPR may also be tried in primary cells [36].

One concern with growing cells from tumor tissue is the potential for contamination with normal cells. To address this, we list here the significant differences we have seen between the normal, benign and cancer cultures that we grow. The cancer-derived cultures maintain the prostate cancer-associated TMPRSS2/ERG fusion [5], which was also observed in the $CD133^{+}/\alpha_2\beta_1integrin^{hi}$ cells. High Gleason grade cancer cultures were also capable of invasion and displayed differential gene expression patterns compared to low Gleason grade and BPH-derived cultures in microarray analysis [11, 13, 37]. There was also differential gene expression between SCs derived from BPH cultures versus SCs derived from cancer-cultures [13]. Of note, these observations were all made when the cultures were at very low passages.

Patient tissue material (normal, benign, and malignant) is an invaluable resource that requires a true collaboration between scientists, surgeons, and patients. If we are to strive to produce relevant data for new therapies, then more laboratories need to work with cells derived from patient tissue and this model should result in more successful therapies being derived.

2 Materials

2.1 Harvesting and Processing Tissue

1. Transport Media: RPMI Medium 1640 plus 10% fetal calf serum, 2 mM L-glutamine, and 1% ABM (antibiotic/antimycotic).
2. Labels: Brady labeller (LabXpert™).
3. Needle biopsies: TruCore II biopsy Instrument 14 g×10 cm, AngioTech.
4. Decontamination: Virusolve, Cairn Technologies, Sheffield.

2.2 Patient Blood

1. Tubes for collecting blood: BD Vacutainer® Plastic, K3EDTA tube with Lavender BD Hemogard™ Closure.
2. Histopaque, Sigma.
3. Qiagen DNeasy blood and tissue kit.
4. Genotyping lymphocytes: Promega Powerplex 16.

2.3 Preserving Tissue: OCT/Formalin

1. OCT: optimum cutter temperature compound, Tissue-Tek, Sakura Finetek Europe BV.
2. Formalin: 10% formaldehyde in PBS.

2.4 Processing and Culturing Primary Prostate Epithelial Cells

1. Syringes: BD Plastipak 10 ml syringe.
2. Blunt needles: Monoject Blunt Cannula 21G.
3. R10: RPMI 1640, 10% FCS, 2 mM L-glutamine, and 1% ABM (antibiotic/antimycotic).
4. SCM (stem cell media): keratinocyte serum-free medium with epidermal growth factor, bovine pituitary extract, 2 ng/ml leukemia inhibitory factor (Millipore), 1 ng/ml GM-CSF (Miltenyi Biotec), 2 ng/ml stem cell factor (First Link), 100 ng/ml cholera toxin (Sigma).
5. Collagenase Solution: Weigh out fresh collagenase (Worthington Biochemical Corporation, Lorne Laboratories Ltd.) with a final concentration of 200 IU/ml in a final volume of 7.5 ml/g of tissue. Dissolve the collagenase in 2.5 ml KSFM and 5 ml of R10. Filter this solution into a sterile container using a 10 ml syringe and 0.2 μM filter.
6. Sterile disposable scalpels: Swann-Morton, size 21.

7. Sterile metal tweezers (oven-treated).
8. Decontamination: Virusolve, Cairn Technologies, Sheffield.
9. Trypsin: 0.05% trypsin and 1 mM EDTA in PBS.
10. Orbital shaker: MaxQMini 4000, Barnstead/Lab-line E-CLASS.
11. STO feeder cells (mouse embryo fibroblast line): STO cells were cultured in R10 medium. *See* ref. 29 for description of how to prepare STOs.
12. Type I collagen-coated petri dishes, BD Biocoat™.
13. Centrifuge: Thermo Electron Corporation IEC CL30R.

2.5 Selection of Cell Subpopulations

2.5.1 From Cultures of Cells

1. BSA Blocking Solution: 0.3% BSA (bovine serum albumin) dissolved in PBS and heat-inactivated at 80 °C for 5 min, filtered through Corning 500 ml filter system.
2. MACS Buffer: 0.5% BSA dissolved in PBS, pH 7.2, plus 2 mM EDTA.
3. MS columns, Miltenyi Biotec.
4. CD133 microbead kit, Miltenyi Biotec.
5. MACSmix Tube Rotator, Miltenyi Biotec.
6. MACS MultiStand, Miltenyi Biotec.
7. OctoMACS Separator, Miltenyi Biotec.
8. BD Falcon 5 ml polypropylene round-bottom tubes.
9. Type I collagen-coated 100 mm petri dishes (BD Biocoat™).

2.5.2 Directly from Tissue

1. Cell strainer (40 μM), BD Falcon.
2. CD24 microbead kit (Miltenyi Biotec).
3. CD31 microbead kit or CD31-biotin antibody (Miltenyi Biotec).
4. Lineage depletion kit human (Miltenyi Biotec).
5. LS columns (Miltenyi Biotec).
6. MS columns (Miltenyi Biotec).
7. Lymphocyte separation medium (LSM) (MP Biomedicals).
8. MACS MultiStand (Miltenyi Biotec).
9. QuadroMACS Separator (Miltenyi Biotec) OR MidiMACS Separator (Miltenyi Biotec).
10. OctoMACS Separator (Miltenyi Biotec) OR MiniMACS Separator (Miltenyi Biotec).
11. MACSmix Tube Rotator (Miltenyi Biotec).

2.6 Differentiation of Cultured Cells, Stromal Cocultures, and Spheroid Growth

1. DH10 media: 50:50 (v/v) DMEM (Dulbecco's Modified Eagle Medium): Hams F7 plus 10% serum.
2. DHT 10 nM 5a-dihydrotestosterone (5a-DHT) or R1881 (also called methyltrienolone—a synthetic androgen.

3. Cell inserts (Falcon).
4. 24-well plates or 12-well plates.
5. Matrigel: growth factor reduced phenol red free Matrigel, BD Bioscience.
6. Non-adherent plates (Nalge Nunc, Tokyo, Japan).

3 Methods

3.1 Harvesting Tissue

In order to obtain patient tissue (*see* **Note 1**), collaboration with a surgeon(s) and pathologist must be initiated and maintained. There are ethical procedures that are required in order to be allowed to obtain, use and store human tissue. Patients have to be invited to voluntarily donate tissue and complete a consent form, and all samples are anonymized. The time and effort to achieve this is often underestimated but there are severe penalties in many countries for the use of ethically unapproved human tissues.

Prostate tissue can be obtained from transurethral resection of the prostate (TURP) (Fig. 1a (i)) or radical prostatectomy operations (Fig. 1a (ii)). TURP procedures are carried out to unblock the urethra due to overgrowth of the prostate, which occurs during benign prostatic hyperplasia (BPH) or as palliative care for a non-resectable tumor, often castration resistant. "Chips" of tissue are obtained, that are removed using a heated wire loop. Malignant tissue can also be obtained from patients undergoing a radical prostatectomy, removal of the whole prostate. The majority of the prostate is required for diagnosis by a pathologist. However, targeted needle biopsies to remove a small piece of cancer tissue and adjacent normal tissue can be carried out without compromising diagnosis (14 gauge needle). Tissue collection can be carried out by the surgeon, research nurse, or dedicated tissue procurement officer. Once sampled, tissue is placed in universal tubes containing transport medium. Tissue processing should happen on the day of the surgery, although successful primary cell cultures have been established following storage of tissue overnight at 4 °C. All tubes containing patient tissue or blood must be labeled with patient code, date, and user (*see* **Note 2**).

3.2 Patient Blood

Blood is collected in heparinized vacutainers and plasma and lymphocytes subsequently isolated using a lymphocyte separation media (LSM) gradient or equivalent. Both are stored at −80 °C. Lymphocyte DNA is used for genotyping and plasma can serve as a material to test potential biomarkers. Genotyping is necessary after establishment of and serial transplant of patient-derived xenografts or serial passaging of primary cultures to match samples with original material.

Lymphocyte DNA extraction is carried out using the Qiagen DNeasy kit according to the manufacturer's instructions.

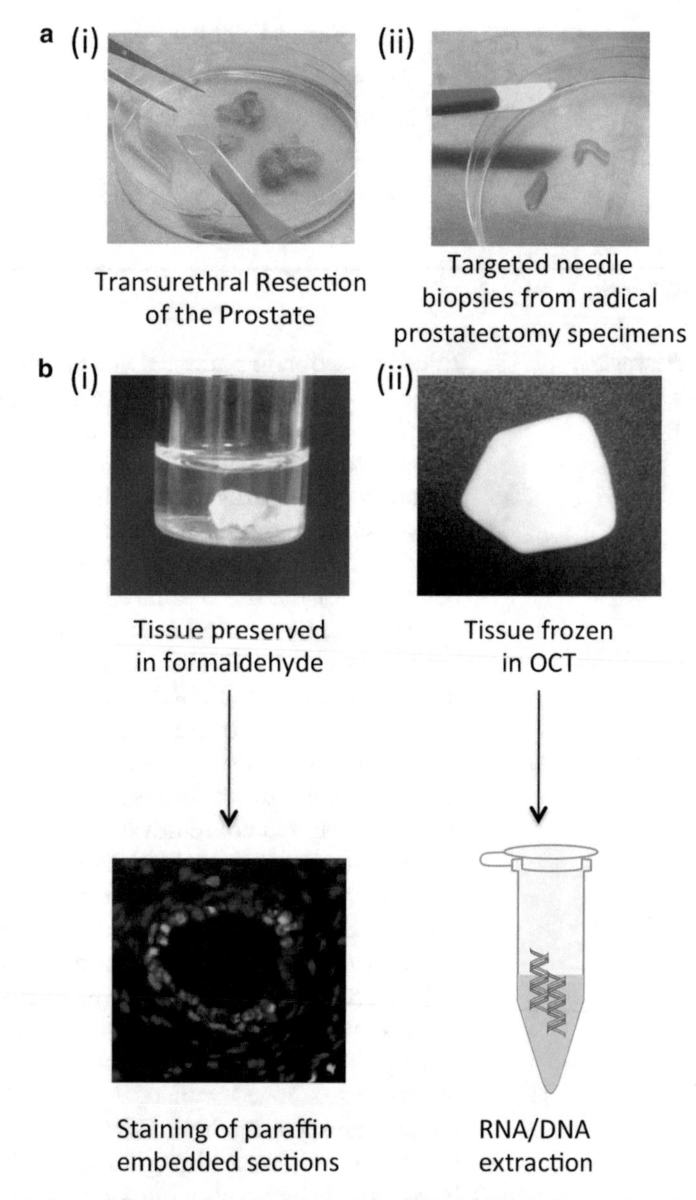

Fig. 1 Uses of tissue taken directly from patient. (**a**) Tissue is acquired from (i) transurethral resection of the prostate (TURP) and (ii) radical prostatectomy procedures. (**b**) Fresh tissue can be preserved in (i) formalin for paraffin-embedding, sectioning, and staining, (ii) OCT, which is suitable for RNA and DNA extraction

Genotyping is then carried out using the Promega Powerplex® 16 System and analyzed using ABI GeneMapper® 4.0.

3.3 Preserving Tissue: OCT/Formalin

Following harvest, tissue pieces are placed in buffered formalin, to then be paraffin-embedded and stained for histology and immunohistochemistry (Fig. 1b (i)) or snap-frozen in OCT for histology (as above) or more commonly for RNA or DNA extraction (Fig. 1b (ii)).

3.4 Processing Tissue and Culturing Primary Prostate Epithelial Cells from Patient Samples

In order to process the tissue on arrival from the hospital, it is helpful to have all materials pre-allocated in the designated laminar flow tissue culture hood, including a beaker of Virusolve disinfectant cleaner at appropriate concentration to collect waste solutions and utensils.

1. Pour the transport media containing tissue into a 10 ml petri dish, aspirate the media without touching the tissue, and rinse in PBS. At this point cut off a piece for storage in OCT and/or formalin (Fig. 1b (i) + (ii)). It is possible to orientate the tissue using tissue marking ink. Label all tubes containing tissue with patient code, date, and initials. To freeze in OCT you need a small plastic dish around 1 cm^2. After placing the tissue in the dish, cover in OCT and flash freeze by lowering it into liquid nitrogen with a ladle. To preserve tissue in formalin, place a piece of the tissue in formalin in a bijou container. Now aspirate the PBS from the rest of the tissue.
2. Pour the collagenase solution on to the remaining tissue and chop up the tissue using a scalpel into ~1 mm × 1 mm pieces. Transfer the collagenase 1 solution containing pieces of tissue into a 125 ml Erlenmeyer flask using a 25 ml pipette and incubate tissue in collagenase at 37 °C overnight in an orbital shaker at 80 rpm.
3. Triturate the digested tissue by repeated pipetting with a 5 ml pipette followed by a syringe and blunt needle. This will break up the clumps but there will still be little pieces floating around containing intact acini.

4a. Decant the mixture into a universal and spin at 380 × *g* for 5 min to sediment the cells. Discard the supernatant using a Pasteur pipette and place it in a waste beaker containing Virusolve disinfectant. Resuspend cells in 10 ml of PBS to wash out collagenase, centrifuge at 380 × *g* for 5 min. Remove the supernatant and repeat this PBS wash once more. Depending on the sample you may need to increase the speed and time to 670 × *g* and 10 min.

4b. If stroma is required from the sample, follow **steps 5–7**. These steps are to separate epithelial acini from stroma using differential centrifugation. For cancer samples (where there are less or no intact acini), or if stroma is not required proceed to **step 9**.

5. Resuspend the segregated cell pellet in 10 ml R10 and centrifuge at 100 × *g* for 1 min. Epithelial acini will settle to the bottom and the supernatant will contain stroma.
6. Collect the acini with a Pasteur pipette, rinsed first with PBS. Hoover up the acini from the bottom of the tube and place them in a sterile labeled universal tube.
7. Repeat 5+6 with the supernatant until all acini are collected (can be 4–5 times). The supernatant remaining is the fibro-

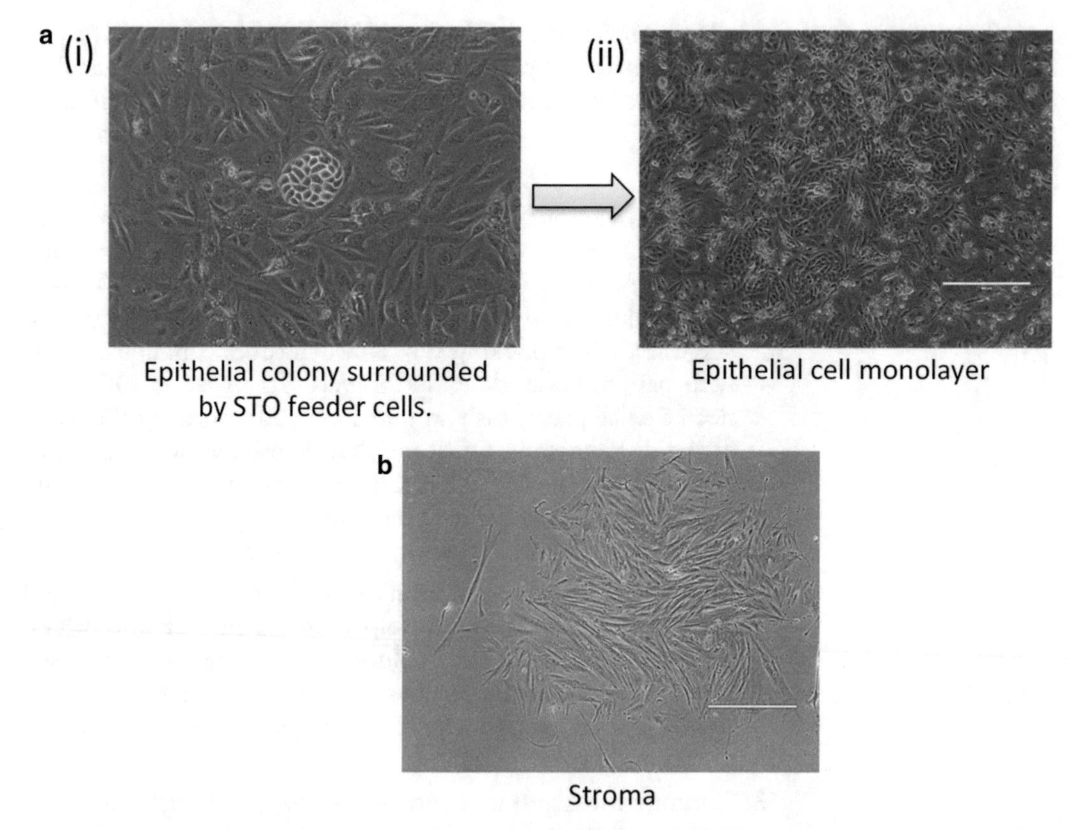

Fig. 2 Culture of primary epithelial and stromal cells. (**a**) (i) Primary prostate epithelial cells grow as a colony on type I collagen coated plates with STO feeder cells. (ii) The cells gradually grow as a monolayer and the STO feeder cells die and detach. (**b**) Stroma can also be grown from the patient tissue after differential centrifugation to separate it from the epithelial acini

blast/stromal component, which can be spun down and resuspended in 12 ml of R10 and placed in a T75 flask for propagation (Fig. 2b).

8. Wash the acini with 10 ml PBS to remove FCS. Aspirate the PBS very carefully with a Pasteur pipette so that you don't disturb the pellet (*see* **Note 3**).
9. Resuspend the acini in 1× trypsin (10 ml) and incubate at 37 °C for 30 min in orbital shaker at 80 rpm. After this incubation, add 10 ml R10 and shake vigorously. Centrifuge at 380 ×g for 5 min and resuspend the pellet in 5 ml PBS. Spin again at 380 ×g for 5 min. Resuspend the pellet in Stem Cell Media.
10. The epithelial cells are plated on a type I collagen plate (100 mm^2) in SCM with STO feeder cells. The total volume should be 5–7 ml. Add more STOs the following day if there is not a confluent layer. Colonies of cells appear within 1 or 2 weeks (Fig. 2a (i)), after which they will grow into a monolayer (Fig. 2a (ii)) (*see* **Note 4**). Cells are maintained by feeding with

Table 1
Examples of antibodies to characterize marker proteins in tissue and cultured cells by immunofluorescence or flow cytometry

Protein	Marker of	Antibody
p63	Basal cells	DAKO, DAK-p63, M7317
Nkx3.1	Luminal cells	Menapath, MP-422-CR01
Pan-cytokeratin	Epithelial cells	Sigma, C2562
CK5	Basal cells	Vector Laboratories, VP-C400
CK18	Luminal cells	Sigma, C8541
AR	Luminal cells	Santa Cruz, sc-816
PSA	Luminal cells	Dako, M0750
PAP	Luminal cells	Dako, M0792
AMACR	Cancer cells	Dako, M3616

SCM every 2 days. STO feeder cells are topped up when necessary to fill the gaps between colonies but without being too crowded. Cells are frozen down at p0 and passaged. Experiments should be carried out at low passages (<p5). Higher passage cells display clear phenotypic changes.

3.5 Characterization of Cultured Samples and Staining of Tissue

To characterize the cultured cells, a panel of antibodies can be used for immunofluorescence or flow cytometry (Table 1). Cultured primary prostate epithelial cells express cytokeratin 5 (Fig. 3a (i)), a typical basal cell marker, with only rare cells expressing cytokeratin 18 (Fig. 3a (ii)), a typical luminal cell marker. Using immunofluorescence, the high and low levels of α2integrin expression can be visualized using the CD49b antibody (Fig. 3a (iii)). Primary cultures also have no or low expression of androgen receptor and PSA suggesting they are not luminal. Cells positive for CD44 and CD24 have been identified in the primary cultures (Fig. 3b). CD44 is considered a basal marker and CD24 is considered a luminal marker. However, lack of other luminal markers and co-expression of CD24 with basal markers suggest that these cells have a basal to intermediate phenotype [38, 39].

Both original and patient-derived xenograft tissue can be stained for prostate-specific, cell-specific, and/or cancer specific markers.

3.6 Selection of Subpopulations of Cells

In order to take into account cellular heterogeneity, it is necessary to select different subpopulations of cells [11]. Sorting of cell subpopulations can be carried out using a FACS machine such as the MoFlo (Beckman Coulter), FACSAria (BD Biosciences), or S3

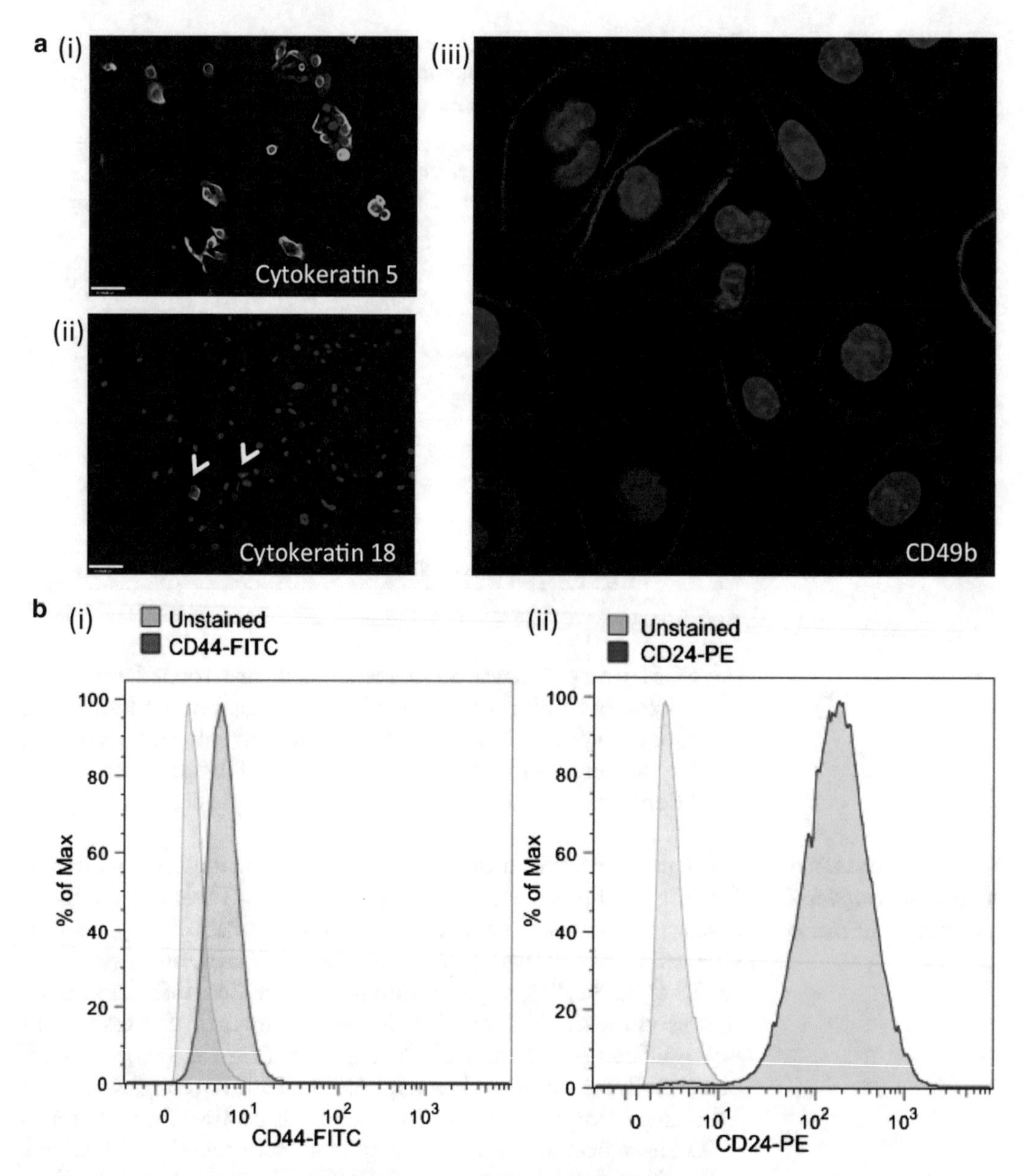

Fig. 3 Characterization of primary prostate epithelial cells. (**a**) When grown on collagen I plates in SCM with STO feeder cells, the primary prostate epithelial cells have a basal phenotype (i) high expression of cytokeratin 5 (ii) low/rare expression of cytokeratin 18 (*white arrowheads*): and (iii) variable expression of α2integrin (CD49b). (**b**) Flow cytometry analysis of basal cells using CD44 and CD24 (ii) antibodies

Cell Sorter (Bio-Rad). An alternative and effective way of enriching for subpopulations is using MACS technology (Miltenyi Biotec) where antibodies linked to magnetic beads bind to cells expressing specific surface receptors. These cells are sorted using columns and a strong magnet. It is the latter method that is described here.

3.6.1 Selection of Subpopulations of Cells: From Cultured Cells

Although the cultured cells are predominantly basal, they are still a heterogeneous mixture and in order to select three different populations of cells from the primary prostate epithelial cultures, there is a two-step selection process. This will result in selecting for α2β1integrin^{lo} (committed basal cells), α2β1integrin^{hi}/CD133^{-} (transit amplifying cells), and α2β1integrin^{hi}/CD133^{+} (stem cells) (Fig. 4a).

1. Type I collagen plates are used for selection of cells using rapid adherence. The first step is to block the collagen I plates using 3 ml of BSA Solution for 1 h at 37 °C.
2. While plates are blocking, cells are trypsinized and resuspended in R10. Following centrifugation at 380 × *g* for 5 min, cells are resuspended in 3 ml of SCM per blocking plate. One blocking plate can be used per plate or per two plates of cells.
3. To collect the α2β1integrin^{hi} cells, incubate for 5–20 min at 37 °C [11, 40–42]. Rapid adhesion to collagen for a 5 min incubation time is a more stringent selection for basal cells than 20 min. Collect the non-adherent cells by vigorously washing in PBS until all the non-adherent fraction is collected. Check using the microscope. Immediately add the fraction to a new blocked plate and incubate for a further 20–35 min. The cells not adhering to the second plate after this (40 min total) will be the committed basal cells (α2β1integrin^{lo}).
4. Collect the rapidly adherent cells (α2β1integrin^{hi} fraction) by incubating with 1× trypsin. After 10 min, collect the fraction in R10 and wash plate vigorously with PBS and pool together. To collect any remaining cells, incubate with 10× trypsin for a further 10 min. Resuspend in R10 and add to the previous cells.
5. Pellet the α2β1integrin^{hi} cells and resuspend in 300 μl MACS buffer and add to a 5 ml round-bottomed tube, then add 100 μl FcR blocking reagent and 100 μl of CD133 beads (final vol 500 μl). Mix well and incubate at 4 °C for 30 min in the tube rotator (*see* **Note 5**).
6. Wash cells once by adding 3 ml of MACS buffer and centrifuge at 380 × *g* for 10 min. Remove supernatant.
7. Resuspend cell pellet in 500 μl MACS buffer and proceed to magnetic separation.
8. To the Miltenyi Biotec MS column add 500 μl of MACS buffer and let buffer run through. Add cells labeled with magnetic beads to column. Columns accept a maximum of 10^8 cells/500 μl buffer per column. However, primary cells are both sticky and large, and there is the risk of them blocking the columns. So typically more than one column is used per sample to minimize blocking, even when starting with 10^7 cells.

9. Wash the column by adding three washes of 500 μl MACS buffer to the reservoir and collect as "CD133 negative fraction."
10. Add 1 ml of MACS buffer to reservoir and firmly flush out cells with plunger into a new tube. The force of the plunger removes the cells from the beads. This is the "CD133$^+$ fraction."
11. Repeat the process by preparing a fresh column then add the CD133$^+$ sample to the column: carry out three washes and collect the fractions. This double-column procedure enriches the population of CD133$^+$ cells from around 70–75% first time round to closer to 95% second time round.
12. Spin down the 1 ml of CD133$^+$ cells (in MACs buffer) at 670×*g* for 5 min. Carefully remove the supernatant and resuspend the cell pellet (which may not be visible) in 200 μl of SCM. Use 10 μl of this to count on a hemocytometer mixed 1:1 with trypan blue. Do not count only in the grid. Count the whole area of hemocytometer containing liquid and this will give you the total number of cells in 10 μl. Multiply this number by 19 to get the number of cells you have left in the remaining 190 μl. This is the only way without compromising viability and sterility to get an estimation of the small number of CD133$^+$/α2β1integrin^{hi} cells (Table 2).
13. You can now freeze the cells or use them in a variety of assays. The low frequency of stem cells means that certain assays designed or adapted for small cell numbers are required. Assays that have been successfully carried out with the CD133$^+$ population include: immunofluorescence, RNA extraction for microarray analysis/qRT-PCR, comet assays, clonogenic assays, ATP viability assays, DNA extraction and DNA methylation assays, telomerase assays, and FISH.

3.6.2 Selection of Subpopulations of Cells: Directly from Tissue (CD24 Luminal Cells)

In order to select luminal (terminally differentiated secretory) cells from the prostate, you have to use fresh tissue, because luminal cells do not grow in culture, since they are terminally differentiated (Fig. 4b).

Follow the protocol from Subheading 3.1 and 3.7, but supplement all of the reagents with 10 nM R1881 to preserve the viability of the luminal cells. After collagenase treatment and trypsin digest, followed by stroma/epithelial acini separation:

1. Resuspend the epithelial fraction in R10 with R1881 and triturate with a blunt needle.
2. Put suspension through a cell sieve to remove any clumps of cells. This should be done with some force.
3. If there is a lot of debris (usually from TURPs but not needle biopsies) then make the volume up to 15 ml and slowly add 15 ml of lymphocyte separating medium (LSM). Add the LSM at the bottom of the tube below the media and cells.

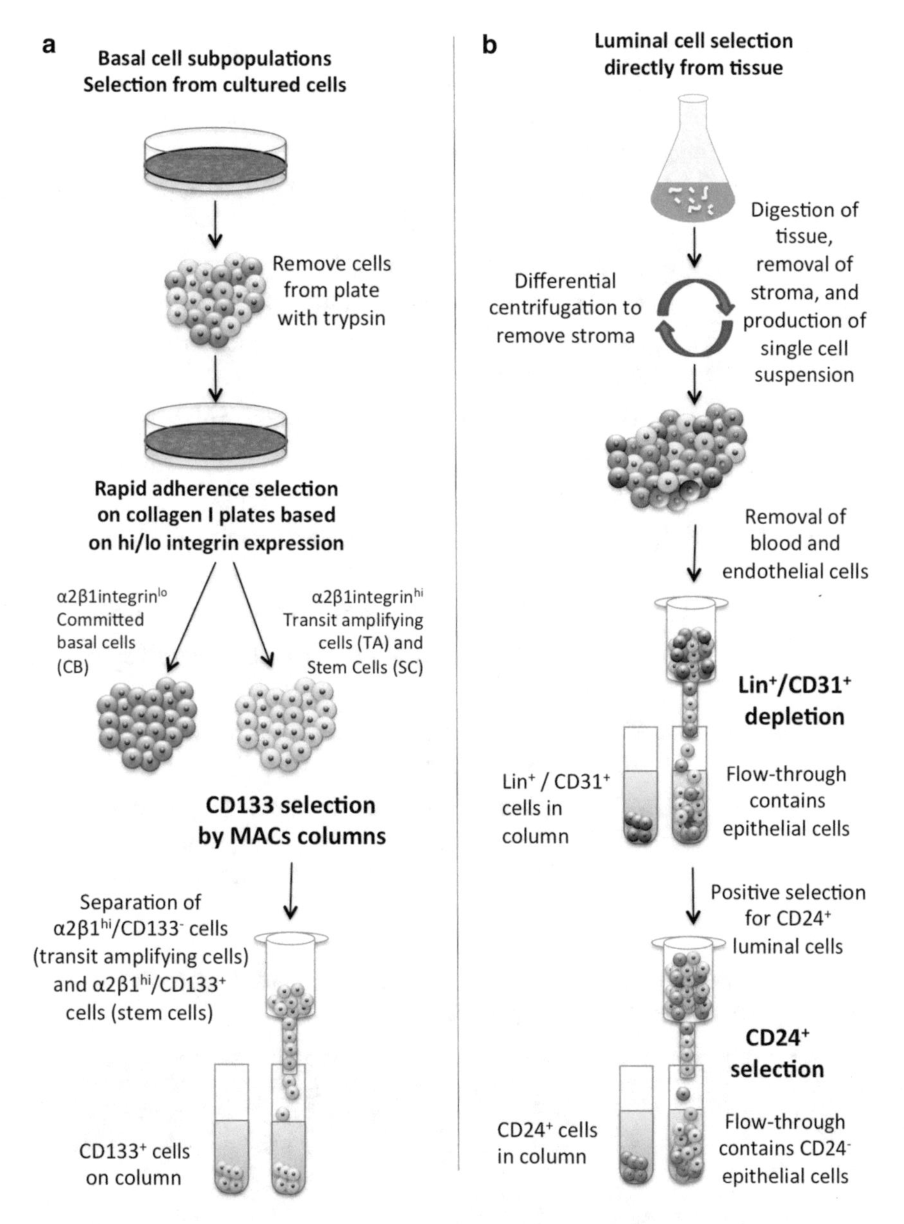

Fig. 4 Selection of subpopulations of cells from cultured cells and directly from patient tissue. Subpopulations of cells can be selected using different surface markers. (**a**) From basal cell cultures, committed basal ($\alpha 2\beta 1^{lo}$), transit amplifying ($\alpha 2\beta 1^{hi}$/CD133^{-}), and stem cells ($\alpha 2\beta 1^{hi}$/CD133^{+}) can be selected. First the CB cells are separated from the TA/SC cells using rapid adherence on collagen I plates. TA/SC cells rapidly adhere. SCs are separated from TA cells using a CD133 antibody attached to a magnetic bead and MACs columns. (**b**) Since luminal cells do not grow in culture, luminal cells have to be selected directly from tissue. Following collagenase and trypsin and differential centrifugation to remove stroma, the cells undergo a Lin/CD31 depletion to remove endothelial and blood cells. The prostate epithelial cells then are labeled with CD24 to positively select for luminal cells

4. Spin at 530 × *g* for 30 min at room temperature with no brakes. Dead cells, ECM fragments, and red blood cells will pellet at the bottom and live cells will remain at the interface between the media and the LSM.
5. Collect this layer of live cells using a Pasteur pipette and transfer to a 50 ml Falcon tube.
6. Add an equal volume of R10 and pellet cells at 530 × *g* for 15 min at 4 °C with brake on. The live epithelial cells will pellet.
7. Remove the supernatant and resuspend the cells in 4 ml of cold MACs buffer.
8. Transfer to a 5 ml polypropylene tube. Spin at 380 × *g* for 5 min at 4 °C.
9. Resuspend cells in 80 μl MACs buffer.
10. Add 20 μl of biotin-antibody cocktail (Lineage depletion kit—human).
11. Incubate for 10 min at 4 °C in the tube rotator.
12. Wash cells by adding 3 ml MACs buffer and spin at 380 × *g* for 5 min.
13. Carefully remove the supernatant and resuspend the cells in 80 μl MACs buffer.
14. Add 40 μl of FcR blocking reagent (CD31 kit), 40 μl of anti-CD31 microbeads (CD31 kit), and 40 μl of anti-biotin microbeads (lineage depletion kit—human) (*see* **Note 6**).

Table 2
Several examples of cultured cells from patient samples indicating starting number of cells in culture prior to selection and number of $CD133^+/\alpha 2\beta 1integrin^{hi}$ cells

Sample	Pathology	Total number of cells	Total number of stem cells	% SCs	Passage No.
A	BPH	1.12×10^7	1623	0.015	4
B	BPH	1.84×10^7	1487	0.008	3
Ci	Normal	2.41×10^7	2214	0.009	3
Cii	Normal	1.45×10^7	181	0.001	4
Di	Normal	1.23×10^7	400	0.003	4
Dii	Normal	1.96×10^7	1620	0.008	4
Diii	Normal	2.16×10^7	576	0.003	2
Ei	Ca 3 + 4 in 10% + fin	1.08×10^7	441	0.004	4
Eii	Ca 3 + 4 in 10% + fin	1.93×10^7	8209	0.042	3
F	Ca 3 + 3 in 1%	2.06×10^7	5753	0.028	3

15. Incubate for 15 min at 4 °C in the tube rotator.
16. Prepare stand, magnet and LS columns. Activate LS columns by adding 3 ml of MACS buffer to the column/Discard the flow-through.
17. Add 4 ml of MACS buffer to wash the cells, then spin at 380 × *g* for 5 min 4 °C.
18. Discard the supernatant (being careful not to aspirate the pellet) and resuspend the cells in 500 μl of MACS buffer and add to the column. The flow-through contains Lin−/CD31− cells. Lin+ and CD31+ cells remain on the column.
19. Wash the columns with three washes of 3 ml of MACS buffer.
20. Collect all the flow-through (Lin−/CD31− fraction) in a universal container and spin at 380 × *g* for 5 min 4 °C. Discard the supernatant.
21. Prepare another column to reselect the Lin−/CD31− fraction. Resuspend the cells in 500 μl of MACS buffer and add to the pre-washed column.
22. Wash the columns 3× with 3 ml of MACS buffer.
23. Collect all the flow-through (Lin−/CD31− fraction) in a universal and spin at 380 × *g* for 5 min 4 °C. Discard the supernatant.
24. Carefully transfer the Lin−/CD31− fraction into a round bottom (4 ml) polypropylene tube in 80 μl of MACs buffer.
25. Add 20 μl of biotin-CD24 antibody and incubate for 15 min at 4 °C in the tube rotator.
26. Wash cells with 4 ml of MACS buffer.
27. Spin at 380 × *g* for 5 min 4 °C and discard the supernatant (being careful not to aspirate the pellet).
28. Resuspend cells in 80 μl of MACS buffer and add 20 μl of anti-biotin Microbeads.
29. Prepare stand, magnet and MS columns. Activate MS columns by adding 500 μl of MACs buffer to the column. Discard the flow-through.
30. Add 3 ml of MACS buffer to the cells and spin 380 × *g* for 5 min 4 °C/Discard the supernatant.
31. Resuspend the cells in 500 μl of MACS buffer and add to the column (Flow-through = $CD24^-$/In the column = $CD24^+$)
32. Wash the columns 3× with 500 μl of MACS buffer.
33. Prepare a new MS column. Activate MS column by adding 500 μl of MACs buffer to the column. Discard the flow-through.
34. Collect the $CD24^+$ fraction by firmly flushing out the cells with the plunger (with 1 ml of MACs Buffer). You can flush it directly into the new pre-washed column.

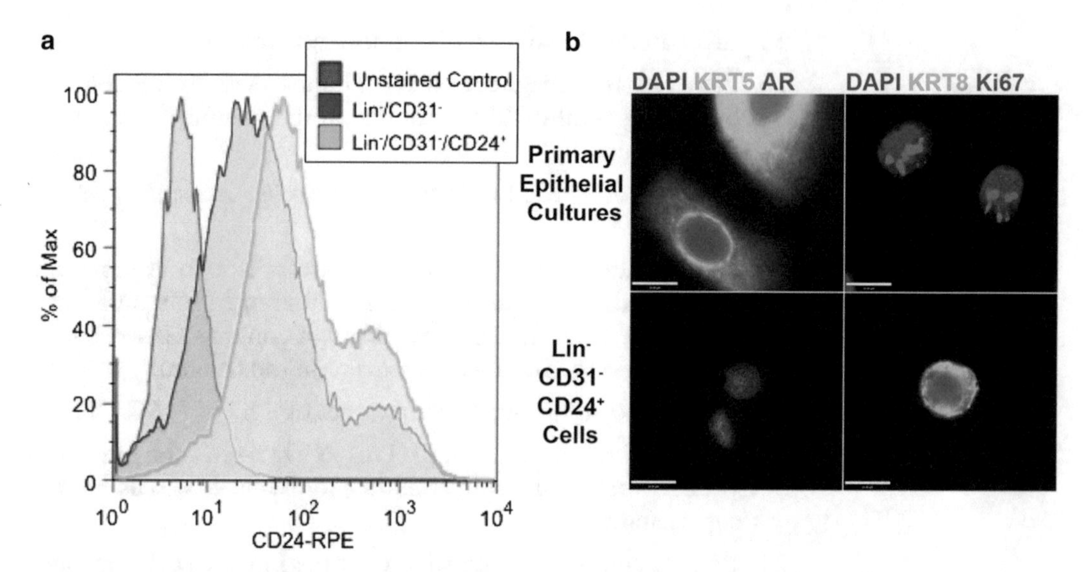

Fig. 5 Characterization of luminal cells selected directly from tissue. (**a**) Flow cytometry showing CD24 expression in luminal cells selected directly from tissue. (**b**) Immunofluorescence staining of selected markers (keratin 5 (KRT5), androgen receptor (AR), keratin 8 (KRT8), and Ki67) on primary epithelial cultures and luminal cells selected directly from tissue

35. Wash the columns 3× with 500 μl of MACS buffer.
36. Collect the CD24+ fraction by firmly flushing out the cells with the plunger (with 1 ml of MACs Buffer).

You now have luminal cells directly from tissue that you can use for immunofluorescence (following cytospin) or RNA or DNA extraction (Fig. 4b). Flow cytometry analysis shows an increase in CD24 expression in the selected population (Fig. 5a). A typical cell yield might be from tens of thousands to hundreds of thousands of cells. These luminal cells are androgen receptor positive and cytokeratin 8 positive [8]. Cells that have been selected can be analyzed by immunofluorescence following preparation by cytospin. In comparison to primary cultures that are AR−, keratin 5+, and Ki67+, the luminal cells are AR+, keratin 5−, and Ki67− (Fig. 5b).

3.7 Differentiation of Cultured Cells, Stromal Cocultures, and Spheroid Growth

Even though the primary prostate epithelial cultures have a predominantly basal phenotype, it is possible to model basal to luminal differentiation with these cells. The epithelial cell differentiation hierarchy [37] is of importance in both normal prostate and prostate disease. The hierarchy of stem cells, basal cells and luminal cells is represented in cancer but with different ratios of cells. Normal and benign human prostate tissues have roughly equal numbers of basal and luminal cells and cancer is defined as an absence of basal cells [37, 43–46]. Importantly, this means that some traits that appear cancer-associated are not cancer-specific and actually translate to differentiation-associated traits [7, 8, 15, 43].

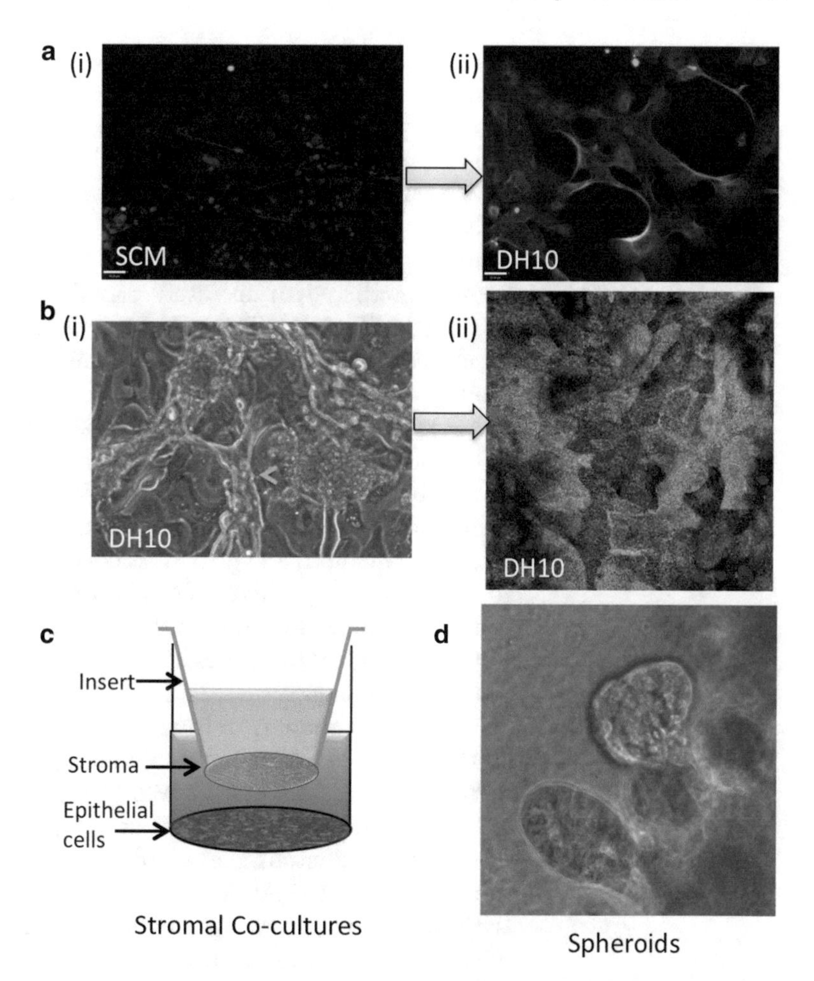

Fig. 6 Differentiation of primary prostate epithelial cells in culture. (**a**) (i) – (ii) When moved from SCM media to DH10 media the primary epithelial cells change in morphology and begin to differentiate (*red*—cytokeratin 1/5/10/14, *green*—cytokeratin 18). (**b**) (i) When grown to confluency in SCM, changed to DH10 with additional DHT and a stromal insert, they differentiate further and even form a bilayer with basal cells on the bottom (*red arrowhead*) and cells with a luminal phenotype on top (*green arrowhead*). (ii) Bilayer indicated by cytokeratin 1/5/10/14 basal layer (*red*) and cytokeratin 18 luminal layer (*green*), nuclei—DAPI stained. (**c**) Stroma and primary epithelial cells can be grown together as a coculture using an insert for one cell type and growing the other in the plasticware. (**d**) Primary epithelial cells can be grown as 3D spheroid cultures

It is possible to differentiate the primary basal cell cultures and drive them towards a luminal phenotype. Culturing the basal cultures to confluency initiates differentiation (Fig. 6a (i)). In addition, changing the cells from stem cell media to DH10 media results in pronounced differentiation over 5–10 days (Fig. 6a (ii)). The monolayer then becomes a bilayer with the upper luminal layer having an elongated cell morphology and the lower compact basal cell layer having a more rounded and compact phenotype (Fig. 6b (i)). The basal cells express cytokeratin 5 and the luminal cells express cytokeratin 18 (Fig. 6b (ii)). In addition the luminal layer cells express rare PSA expressing cells [10, 42].

In order to push cells to a more fully differentiated phenotype, inductive stromal cocultures can be added. This involves growing primary prostate epithelial cells on collagen I plates in stem cell media together with an insert containing stroma cells grown in R10 media (Fig. 6c). The arrangement can also be reversed if necessary with the stroma on the plate and the epithelial cells in the insert.

Stromal cocultures are also used when growing cells in 3D. Three-dimensional models of prostate cell lines have been grown using a variety of methods, including Matrigel and non-adherent plates. Depending on the starting cell type, the additives to the cell medium and growth as cocultures by addition of stroma to cell inserts, the spheroids can be solid or hollow [47–52]. They can also show evidence of different stages of cell differentiation. For primary prostate epithelial cells, spheroids are formed in a similar way (Fig. 6d).

For non-adherent culture, freshly isolated primary prostate epithelial cells are plated in non-adherent plates in SCM (100 μl in a 96-well plate, 2 ml in a 12-well plate).

For culture in Matrigel, cells are plated in 50 μl volume in 96-well non-adherent plates. After 1 week, to allow cell aggregation, 50 μl of 8% (v/v) Matrigel (growth factor-reduced) is added to the cells (final percentage of 4% Matrigel). Often, a 50% Matrigel plug can be placed first in the well to prevent invasive cells from adhering to the plastic on the bottom of the well. Matrigel is stored at −20 °C and thawed on ice at 4 °C. Higher temperatures lead to gel formation, so all tubes, pipette tips, and anything being used to aliquot Matrigel are kept in the fridge.

4 Notes

1. Local and national laws with regard to ethical procedures and approvals should be observed. All agreements should be in place prior to retrieval of patient material.
2. It is wise to invest in a professional labeling system to ensure clear labeling for long-term storage. This should be done in conjunction with a clear and up-to-date database to record location and destination of all samples.
3. If a culture of primary prostate epithelial cells is required, without the need for further selection of subpopulations, explants can be grown on standard tissue culture plasticware in keratinocyte serum-free medium (KSFM) with epidermal growth

factor, bovine pituitary extract, and 2 mM glutamine (no additional supplements) and without STO feeder cells. The acini are plated after the overnight collagenase digestion and before the trypsin digest.

4. When waiting for prostate epithelial cells to grow, some patience is required.
5. When incubating cells with antibodies angle the tube such that the liquid is mixing well but not going into the lid of the tube. This is so that no cells are lost because they are stuck up the side of the tube or inside the lid.
6. This protocol was developed prior to the availability of anti-biotin CD31antibody, so now it is possible to do only one antibody incubation step (**step 10**) with 20 μl of biotin-antibody cocktail (Lineage depletion kit—human) plus 5 μl of anti-biotin CD31, followed by the incubation with anti-biotin microbeads (**step 14**).

Acknowledgements

This work was supported by a Yorkshire Cancer Research Core Grant (Y257PA). Photo credits to: Paula Kroon, Rachel Adamson, Shona Lang, Paul Berry, Stephanie Swift.

References

1. Chaproniere DM, McKeehan WL (1986) Serial culture of single adult human prostatic epithelial cells in serum-free medium containing low calcium and a new growth factor from bovine brain. Cancer Res 46(2):819–824
2. McKeehan WL, Adams PS, Rosser MP (1984) Direct mitogenic effects of insulin, epidermal growth factor, glucocorticoid, cholera toxin, unknown pituitary factors and possibly prolactin, but not androgen, on normal rat prostate epithelial cells in serum-free, primary cell culture. Cancer Res 44(5):1998–2010
3. Frame FM et al (2010) Development and limitations of lentivirus vectors as tools for tracking differentiation in prostate epithelial cells. Exp Cell Res 316(19):3161–3171
4. Hager S et al (2008) An internal polyadenylation signal substantially increases expression levels of lentivirus-delivered transgenes but has the potential to reduce viral titer in a promoter-dependent manner. Hum Gene Ther 19(8): 840–850
5. Polson ES et al (2013) Monoallelic expression of TMPRSS2/ERG in prostate cancer stem cells. Nat Commun 4:1623
6. Kroon P et al (2013) JAK-STAT blockade inhibits tumor initiation and clonogenic recovery of prostate cancer stem-like cells. Cancer Res 73(16):5288–5298
7. Oldridge EE et al (2013) Retinoic acid represses invasion and stem cell phenotype by induction of the metastasis suppressors RARRES1 and LXN. Oncogenesis 2:e45
8. Pellacani D et al (2014) DNA hypermethylation in prostate cancer is a consequence of aberrant epithelial differentiation and hyperproliferation. Cell Death Differ 21(5): 761–773
9. Pellacani D et al (2011) Regulation of the stem cell marker CD133 is independent of promoter hypermethylation in human epithelial differentiation and cancer. Mol Cancer 10:94
10. Swift SL, Burns JE, Maitland NJ (2010) Altered expression of neurotensin receptors is associated with the differentiation state of prostate cancer. Cancer Res 70(1):347–356
11. Collins AT et al (2005) Prospective identification of tumorigenic prostate cancer stem cells. Cancer Res 65(23):10946–10951

12. Adamson RE et al (2012) In vitro primary cell culture as a physiologically relevant method for preclinical testing of human oncolytic adenovirus. Hum Gene Ther 23(2):218–230
13. Birnie R et al (2008) Gene expression profiling of human prostate cancer stem cells reveals a pro-inflammatory phenotype and the importance of extracellular matrix interactions. Genome Biol 9(5):R83
14. Frame FM et al (2013) HDAC inhibitor confers radiosensitivity to prostate stem-like cells. Br J Cancer 109(12):3023–3033
15. Rane JK et al (2014) Conserved two-step regulatory mechanism of human epithelial differentiation. Stem Cell Rep 2(2):180–188
16. Guo C et al (2012) Epcam, CD44, and CD49f distinguish sphere-forming human prostate basal cells from a subpopulation with predominant tubule initiation capability. PLoS One 7(4):e34219
17. Williamson SC et al (2012) Human alpha(2) beta(1)(HI) CD133(+VE) epithelial prostate stem cells express low levels of active androgen receptor. PLoS One 7(11):e48944
18. Jeter CR et al (2011) NANOG promotes cancer stem cell characteristics and prostate cancer resistance to androgen deprivation. Oncogene 30(36):3833–3845
19. van den Hoogen C et al (2010) High aldehyde dehydrogenase activity identifies tumor-initiating and metastasis-initiating cells in human prostate cancer. Cancer Res 70(12): 5163–5173
20. Garraway IP et al (2010) Human prostate sphere-forming cells represent a subset of basal epithelial cells capable of glandular regeneration in vivo. Prostate 70(5):491–501
21. Buhler P et al (2010) Primary prostate cancer cultures are models for androgen-independent transit amplifying cells. Oncol Rep 23(2): 465–470
22. Jiang M et al (2010) Functional remodeling of benign human prostatic tissues in vivo by spontaneously immortalized progenitor and intermediate cells. Stem Cells 28(2):344–356
23. Guzman-Ramirez N et al (2009) In vitro propagation and characterization of neoplastic stem/progenitor-like cells from human prostate cancer tissue. Prostate 69(15):1683–1693
24. Attard G et al (2009) A novel, spontaneously immortalized, human prostate cancer cell line, Bob, offers a unique model for pre-clinical prostate cancer studies. Prostate 69(14): 1507–1520
25. Goldstein AS et al (2008) Trop2 identifies a subpopulation of murine and human prostate basal cells with stem cell characteristics. Proc Natl Acad Sci U S A 105(52):20882–20887
26. Brown MD et al (2007) Characterization of benign and malignant prostate epithelial Hoechst 33342 side populations. Prostate 67(13):1384–1396
27. Heer R et al (2006) KGF suppresses alpha-2beta1 integrin function and promotes differentiation of the transient amplifying population in human prostatic epithelium. J Cell Sci 119(Pt 7):1416–1424
28. Shepherd CJ et al (2008) Expression profiling of CD133+ and CD133- epithelial cells from human prostate. Prostate 68(9):1007–1024
29. Niranjan B et al (2013) Primary culture and propagation of human prostate epithelial cells. Methods Mol Biol 945:365–382
30. Blow NS (2011) Right cell, wrong cell. BioTechniques 51(2):75
31. Nardone RM (2008) Curbing rampant cross-contamination and misidentification of cell lines. BioTechniques 45(3):221–227
32. Phuchareon J et al (2009) Genetic profiling reveals cross-contamination and misidentification of 6 adenoid cystic carcinoma cell lines: ACC2, ACC3, ACCM, ACCNS, ACCS and CAC2. PLoS One 4(6):e6040
33. Barallon R et al (2010) Recommendation of short tandem repeat profiling for authenticating human cell lines, stem cells, and tissues. In vitro cellular & developmental biology. Animal 46(9):727–732
34. Schenk E et al (2014) Preclinical safety assessment of Ad[I/PPT-E1A], a novel oncolytic adenovirus for prostate cancer. Human gene therapy. Clin Dev 25(1):7–15
35. Ulukaya E et al (2013) Differential cytotoxic activity of a novel palladium-based compound on prostate cell lines, primary prostate epithelial cells and prostate stem cells. PLoS One 8(5):e64278
36. Niu J, Zhang B, Chen H (2014) Applications of TALENs and CRISPR/Cas9 in human cells and their potentials for gene therapy. Mol Biotechnol 56(8):681–688
37. Maitland NJ et al (2010) Prostate cancer stem cells: do they have a basal or luminal phenotype? Horm Cancer 2(1):47–61
38. Hudson DL et al (2001) Epithelial cell differentiation pathways in the human prostate: identification of intermediate phenotypes by keratin expression. J Histochem Cytochem 49(2):271–278
39. Hudson DL et al (2000) Proliferative heterogeneity in the human prostate: evidence for epithelial stem cells. Lab Investig 80(8):1243–1250

40. Collins AT et al (2001) Identification and isolation of human prostate epithelial stem cells based on alpha(2)beta(1)-integrin expression. J Cell Sci 114(Pt 21):3865–3872
41. Richardson GD et al (2004) CD133, a novel marker for human prostatic epithelial stem cells. J Cell Sci 117(Pt 16):3539–3545
42. Robinson EJ, Neal DE, Collins AT (1998) Basal cells are progenitors of luminal cells in primary cultures of differentiating human prostatic epithelium. Prostate 37(3):149–160
43. Rane JK, Pellacani D, Maitland NJ (2012) Advanced prostate cancer--a case for adjuvant differentiation therapy. Nat Rev Urol 9(10):595–602
44. Nagle RB et al (1987) Cytokeratin characterization of human prostatic carcinoma and its derived cell lines. Cancer Res 47(1):281–286
45. Humphrey PA (2007) Diagnosis of adenocarcinoma in prostate needle biopsy tissue. J Clin Pathol 60(1):35–42
46. El-Alfy M et al (2000) Unique features of the basal cells of human prostate epithelium. Microsc Res Tech 51(5):436–446
47. Chambers KF et al (2011) Stroma regulates increased epithelial lateral cell adhesion in 3D culture: a role for actin/cadherin dynamics. PLoS One 6(4):e18796
48. Lang SH et al (2010) Modeling the prostate stem cell niche: an evaluation of stem cell survival and expansion in vitro. Stem Cells Dev 19(4):537–546
49. Lang SH et al (2001) Prostate epithelial cell lines form spheroids with evidence of glandular differentiation in three-dimensional Matrigel cultures. Br J Cancer 85(4):590–599
50. Lang SH et al (2006) Differentiation of prostate epithelial cell cultures by Matrigel/stromal cell glandular reconstruction. In vitro cellular & developmental biology. Animal 42(8–9):273–280
51. Lang SH et al (2001) Experimental prostate epithelial morphogenesis in response to stroma and three-dimensional Matrigel culture. Cell Growth Differ 12(12):631–640
52. Pearson JF et al (2009) Polarized fluid movement and not cell death, creates luminal spaces in adult prostate epithelium. Cell Death Differ 16(3):475–482

Chapter 13

In Vivo Imaging of Nuclear Receptor Transcriptional Activity

D. Alwyn Dart and Charlotte L. Bevan

Abstract

Nuclear receptors drive key processes during development, reproduction, metabolism, and disease. In order to understand and analyze, as well as manipulate, their actions it is imperative that we are able to study them in whole animals and in a spatiotemporal manner. The increasing repertoire of transgenic animals, expressing reporter genes driven by a specific nuclear receptor, enables us to do this. Use of luciferase reporter genes is the method of choice of many researchers as it is well tolerated, relatively easy to use, and robust. Further, luciferase lends itself to the process as it can penetrate tissue and can be manipulated to degrade rapidly thus allowing a dynamic response. However, limited resolution, lack of quantitation, and the largely two-dimensional images acquired make it desirable to support results using ex vivo imaging and enzymatic and/or immunohistochemical analysis of dissected tissue. As well as enabling the visualization of nuclear receptor signaling in wild-type animals, crossing these mouse models with models of disease will provide invaluable information on how such signaling is dysregulated during disease progression, and how we may manipulate nuclear receptor signaling in therapy. The use of in vivo imaging therefore provides the power to determine where and when in development, aging, and disease nuclear receptors are active and how ligands or receptor modulators affect this.

Key words Nuclear hormone receptor, In vivo imaging, Luciferase, Transgenic mice, Hormone, Steroid, Response element

1 Introduction

As nuclear receptors are transcription factors, they lend themselves to the use of reporter genes as a measure of their activity, and such assays have been in widespread use for decades, with clearly defined protocols available (e.g., [1]). Given the many and varied target tissues of most nuclear receptor ligands, it is clearly desirable to be able to study their action, in real time, in a more physiologically relevant, whole animal setting. Reporter genes encode easily detectable protein products and several reporters have been classically used in the in vitro setting including CAT (chloramphenicol acetyltransferase) and LacZ (β-galactosidase). The first transgenic mouse models used a lacZ reporter gene under the control of a specific nuclear receptor, either directly or via GAL4, and enabled staining

Iain J. McEwan (ed.), *The Nuclear Receptor Superfamily: Methods and Protocols*, Methods in Molecular Biology, vol. 1443, DOI 10.1007/978-1-4939-3724-0_13, © Springer Science+Business Media New York 2016

of fixed tissues or embryos, e.g., [2, 3]. However, these are unsuitable for the in vivo setting due to toxicities and the requirement for fixing or cell lysis. Therefore, in vivo imaging involves reporters expressing proteins which are externally detectable, i.e., bioluminescent and fluorescent (optical imaging), or proteins which bind or concentrate radioactive tracers, which emit gamma (and positron) radiation (radionuclide imaging). Here, we deal mainly with bioluminescent reporters as these have been proven to be relatively cheap, user and animal friendly, and luminescence penetrates tissue much more efficiently than fluorescence, so has been used in the development of several nuclear receptor reporter mice. The group of Adriana Maggi were first to develop such a system in the whole animal, and published a ground-breaking study in 2001 demonstrating imagable luciferase expression in a transgenic mouse, which expressed a randomly integrated estrogen-responsive luciferase reporter gene [4]. Subsequent work used this model to follow estrogen activity in different tissues during the menstrual cycle, and also exploited it to demonstrate IGF-1 activation of the estrogen receptor (ER) [5, 6]. Since, transgenic models have been developed for imaging activity of further receptors including PPAR, FXR and AR [7–9]. Our group recently reported an AR-reporter mouse expressing luciferase under control of endogenous AR in a non tissue-restricted manner and it is in this model that our techniques described here have been developed [10].

As well as describing in vivo imaging of the whole animal (*see* **Note 1**), we describe use of ex vivo imaging and enzymatic analysis of dissected tissues to both give more quantitative results and also determine more accurately the exact location from which the light signal is emitted. This will enhance and support results obtained by in vivo imaging considerably. Dependent on the organ emitting luminescence, the signal may be reduced by absorption from fur or tissue (the darker the tissue the greater the absorbance, *see* **Note 2**), also luminescent imaging is essentially two-dimensional (although techniques exist to provide "virtual" 3D images), so pinpointing exact location from such images may be problematic. That said, the use of in vivo imaging is invaluable since it allows longitudinal studies of the same animal hence tracking of signal throughout life or disease progression, it reduces the animal numbers required, and is a noninvasive procedure. Combined with ex vivo imaging and analysis after termination of the experiment, this technique provides the power to determine where and when in development, aging, and disease nuclear receptors are active.

2 Materials

The results obtained will be largely reliant on the use of a suitable reporter construct, so much effort will be expended on optimizing this before imaging begins. The specificity of the reporter gene will be driven by the promoter chosen (*see* **Note 3**). Where a transgene is being randomly integrated into the genome, insulators (or boundary elements) should be considered (*see* **Note 4**). Given the uncertainties associated with random integration, such as position-dependent effects (*see* **Note 5**), disruption of endogenous genes and concatamerization of the transgene resulting in copy number variation, it may be preferable to target the transgene to a specific locus (*see* **Note 6**).

In terms of the reporter gene, luciferase, from the North American firefly (*Photinus pyralis*) is rapidly becoming the standard reporter of choice due to its high specific activity and low background (compared to fluorescence for example). d-luciferin, the substrate, is a low molecular weight, organic benzothiazole-based compound that freely diffuses across cell membranes and passes across the blood–brain, blood–testis, and blood–placental barriers [10, 11]. Importantly, the luciferin substrate is nontoxic, non-metabolized and non-immunogenic. Luciferin is excreted by the kidneys, which makes pharmacokinetics easier to study—as inflow and outflow are simply concentration driven. Additionally important is the availability of luciferase protein tagged with degradation signals (hPEST) which eliminate luciferase protein after reporter gene stimulus has been removed—allowing the reporter system to follow dynamic changes. Other bioluminescent reporter enzymes include clickbeetle red and clickbeetle green luciferase, copepod luciferase (*Gaussia princeps*), and the sea pansy renilla luciferase (*Renilla reniformis*).

Since it is essential to test and optimize the reporter in vitro before undertaking in vivo experiments, we will also address this aspect of imaging.

2.1 In Vitro Testing of Hormone Reporters in Transfected Cell Lines

1. Cell line(s) of choice.
2. Reporter plasmids, e.g., pGl-4 basic luciferase reporter vector from Promega or equivalent with the inserted gene–promoter fusion or constructed minimal promoter.
3. Standard transfection materials, e.g., electroporation or lipid agent.
4. Selection agents, e.g., G418, blasticidin, puromycin, or zeocin (for stable transfection).
5. Luminometer or IVIS-100, 200, Lumina or Spectrum CCD (PerkinElmer) linked to a PC.

2.2 In Vivo Imaging

The methods below are applicable both to transgenic reporter mice and to mice bearing xenografts expressing the reporter gene (*see* **Note 7**).

1. Ligand/xenobiotic of choice, dissolved in solvent, e.g., ethanol or DMSO, and then diluted in suitable carrier for injection, e.g., vegetable or sunflower oil. For oral gavage, use suitable amount of compound in 0.5% methylcellulose (CMC). Different routes of administration are likely to be required for different drugs/

hormones/xenobiotics and this will be influenced by the solubility of the substance, the avoidance of toxicity in the animal, bioavailability etc. The route of administration and duration of treatment prior to imaging will vary from experiment to experiment and so we have not attempted to describe them exhaustively here. Further, it may or may not be desirable/necessary to remove endogenous hormones from the animal by prior gonadectomy or adrenalectomy (*see* **Note 8**). Examples of hormone administration protocols are given in **Note 9**.

2. Needles 21G or lower.
3. IVIS 100, IVIS 200, IVIS Lumina or IVIS Spectrum CCD (PerkinElmer) linked to a PC.
4. Anesthetic system—isofluorane gas or injectable sedative (4:1 mixture of ketamine (100 mg/ml) and xylazine (20 mg/ml)).
5. Luciferin substrate (d-luciferin, potassium salt) 30 mg/ml in injectable saline, filter-sterilized (0.2 μm). Protected from light.
6. Heated pad covered with black sheet of paper.
7. Black plastic sheets to separate individual animals.
8. Recovery facilities, e.g., heated box.
9. Software—e.g., LivingImage.

3 Methods

3.1 In Vitro Testing of Hormone Reporters in Transfected Cell Lines

1. Use cell lines either transiently or stably transfected with the luciferase reporter construct of choice. For stable cell line production consult the vector and selection agent guidelines for times and dosages.
2. For steroid hormone induction studies, cells must be grown in medium with charcoal stripped serum (5 %) for 72 h (approx.) prior to hormone treatment.
3. Allow at least 6–8 h for gene induction and luciferase protein expression.
4. Cells may be imaged live by the addition of luciferin substrate directly to the media (1×), and imaged by IVIS system or luminometer.
5. Cells may be imaged as a monolayer or as cell pellets.
6. Luciferase activity may also be monitored from cell lysates using a luminometer.
7. Place cell plate (any 6- to 96-well or individual dishes) in the IVIS chamber and close door.
8. Image acquisition using IVIS Living Image Software:
   Start the software, and initialize the system and wait for the CCD temperature to drop.

Acquire the image sequence by clicking the acquire button, and/or following the imaging wizard built into the software.

9. Depending on equipment and software in use:
   Confirm the excitation filter is closed and the emission filter is set to open.
   Set the binning level and the F/Stop—either set to default or set to required level.
   Set the field of view (FOV) suitable for your plate size—which can be checked by an initial greyscale image.
   Set the exposure time.
   Click acquire (*see* **Note 10**)
10. Image and data analysis: Open the image of interest. In the region of interest (ROI) tools, select measurement ROI from the drop-down list.
11. Select shape of ROI—circle, square, or grid for 6–96 well plates. ROIs can be adjusted by clicking on the ROI and dragging to the desired shape.
12. If the image is part of a sequence, select "apply to sequence" or select, copy and paste selected ROI to each subsequent image.
13. Once the ROI are drawn click on "measure". This will open a window for the measurements. Ensure you are measuring ROI as photon flux data (photons/s).
14. Cells may be washed in PBS and the medium replaced for continued growth and sequential studies (Fig. 1).

3.2 In Vivo Imaging

The imaging method outlined below has been optimized using IVIS 100, IVIS Spectrum, and IVIS Lumina XR machines but should be applicable to any IVIS system. Colocation with or direct access to a

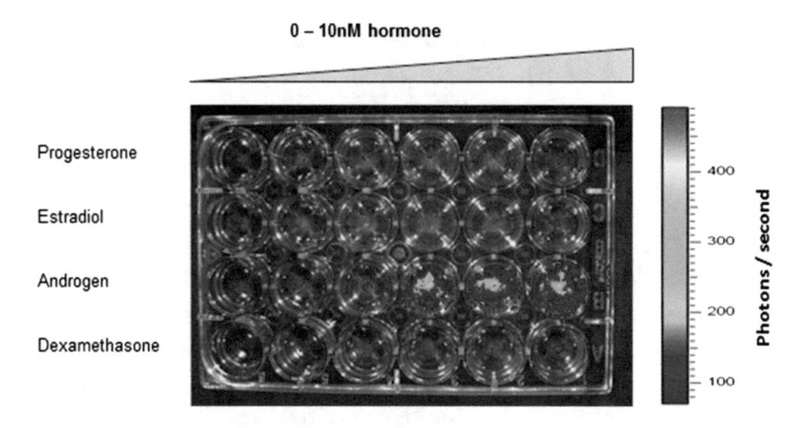

Fig. 1 Example of stably transfected LNCaP/Luc cells containing an androgen receptor luciferase reporter construct. These cells were starved for 72 h and treated with various steroid hormones for 24 h then imaged using IVIS-100 bioluminescent imaging equipment

designated animal handling room is necessary for transfer of the mice from anesthetic apparatus to the machine and back to recovery.

1. Inject 150 mg/kg of luciferin in a volume of 150–200 μl. This may be intraperitoneal, intravenous, or subcutaneous.
   (a) Intraperitoneal (i.p.) injection*:* Select lower left quadrant of the abdomen, with the animal tilted with its head down. Needle should be bevel-side up. Penetrate through the abdominal wall (4–5 mm).
   (b) Intravenous (i.v) injection: Select tail vein (or retro orbital may be used in some circumstances). Gentle warming of the animal will dilate the vein and aid in injection and vein visualization.
   (c) Subcutaneous injection (s.c.): Luciferin substrate is injected into the scruff of skin below the neck on the back of the animal.
2. If anesthetizing using isofluorane gas:
   (a) Weigh the air filter before and after the procedure to ensure personal safety.
   (b) Fill the vaporizer with isofluorane.
   (c) Open the oxygen tank valve and set gas flow to 1 L/min.
   (d) Place animals in an induction chamber and secure lid.
   (e) Open the valve allowing airflow into the induction chamber. Turn the dial on the vaporizer to 3–4%.
   (f) Animals should fall asleep within 2–3 min. Immediately transfer animals to the imaging chamber, and open the valve allowing anesthesia into the chamber nose cones—closing the flow to the induction chamber.
   (g) After 1–2 min, turn the anesthesia down to 1.5–2%.
3. If anesthetizing using injectable ketamine:
   (a) Prepare 4:1 mixture of ketamine (100 mg/ml) and xylazine (20 mg/ml)
   (b) Inject volume 30–50 μl.
   (c) Anesthesia will take effect within 2–3 min. Transfer animals to imaging chamber. Animal should remain sedated for 30 min approx. and will require a warming pad throughout imaging and recovery.
4. For Image acquisition using IVIS Living Image Software—start the software, initialize the system and wait for the CCD temperature to drop (indicated by green light).
5. Place the anesthetized and luciferin-injected mice in the imaging chamber and close the door.

6. Acquire the image sequence by clicking the acquire button, and/or following the imaging wizard built into the software.
7. Depending on equipment and software in use:
   (a) Confirm the excitation filter is set to block/closed and the emission filter is set to open.
   (b) Set the binning level and the F/Stop—either set to default or set to required level.
   (c) Set the field of view (FOV). Wide angle for multiple animals, narrow for single animals or specific region.
   (d) Check animals are all in frame by taking an initial greyscale image.
   (e) Set the exposure time.
   (f) Click acquire.
8. Determine optimum (plateau) luminescence by sequential imaging every 1–2 min continuously. An example of data is given in Fig. 2. Each organ in the body will have its own unique blood supply and clearance rates, and therefore, luciferin substrate levels may vary. Sequential/timecourse imaging will clarify rates of tissue distribution.
9. For image and data analysis: Open the image of interest.
10. In the region of interest (ROI) tools, select measurement ROI from the drop-down list.
11. If the image is part of a sequence—select "apply to sequence" or select, copy and paste selected ROI to each subsequent image.
12. Select shape of ROI—circle, square, or grid. ROIs can be adjusted by clicking on the ROI and dragging to the desired shape.
13. Once the ROI are drawn click on "measure". This will open a window for the measurements. Ensure you are measuring ROI as photon flux data (photons/s).
14. In sequential/timecourse imaging, photon flux measurements can be made at the plateau of luminescence.
15. For multimodality and co-registration of 3D images, *see* **Note 11**.

3.3 Ex Vivo Organ Imaging

1. Animals injected with luciferin substrate should be culled just before peak luciferin circulation (usually 10–20 min post-injection).
2. Specific organs are removed as rapidly as possible, washed quickly in PBS and placed in a plastic dish or plate (a large dish is advantageous, to avoid proximity reflection on plastic), inside the IVIS imager.
3. The field of view must be adjusted for smaller organs.
4. Data acquisition must be carried out rapidly, before organs begin to dry out.

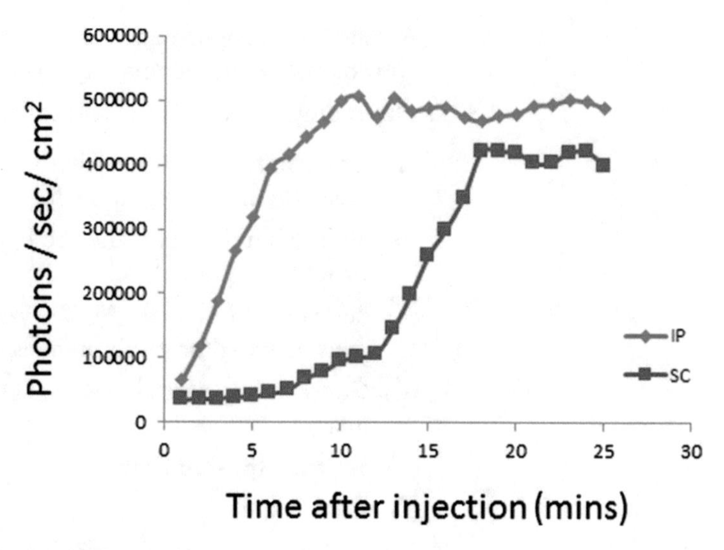

Fig. 2 Luciferase activity in the whole mouse injected with luciferin substrate via intraperitoneal (IP) or subcutaneous (SC) routes and monitored by sequential imaging over time for establishment of plateau phase

5. Organs may be collected for formalin fixed paraffin embedding at this point—for immunostaining or in situ hybridization.

3.4 Luciferase Activity Measurement in Tissue Extracts

1. Tissues must be collected from animals at least 24 h after any prior luciferin injection, to allow for body clearance to prevent downstream contamination of samples.
2. Tissue may be collected and snap-frozen in liquid nitrogen for later measurements. In addition, it may be desirable to collect tissue for other assays, such as RNA extraction for quantitiative PCR.
3. Small tissue pieces are homogenized via several possible routes:
   - (i) Tissue lyser (Qiagen) using lysis buffer and beads.
   - (ii) Tissue homogenizer, using lysis buffer (e.g., reporter assay lysis buffer Promega)
   - (iii) Tissue grinding beads and microfuge pestles.
   - (iv) Sonication in lysis buffer.

   Methods i and ii are best for tissues high in collagen and/or cartilage, which may be difficult to lyse by tissue grinding or sonication.
4. Tissue lysates must be kept on ice at all times. If samples are heated during homogenization, this will impact upon luciferase activity and a method that generates less heat shoudl be used (e.g., microfuge pestles).

5. Lysates are centrifuged briefly to remove larger non-homogenized debris at 4 °C.
6. Lysates are used in a standard luciferase assay. Luciferase activity (photons/s) can be normalized and expressed as activity per mg of protein, determined by standard Bradford assay.

4 Notes

1. Animal experiments must have received ethical approval and be carried out under a valid project license by appropriately trained personnel. Local and national laws and guidelines, e.g., the NCRI guidelines [12] should be followed,
2. Mouse genetic background/coat color should be considered. Black fur, to some extent (estimates vary), absorbs light, and therefore reduces luminescence signal. If this is an issue, dependent on the region the signal comes from, it may be desirable to remove the hair from such mice by shaving or hair removal cream. We have not found this to be necessary in our C57/Bl6 background, possibly due to ventral imaging. Another option is using a white strain such as Balb/c; however, white fur has higher background luminescence and it may be necessary to subtract this, possibly resulting in higher mouse numbers and reduced reliability.
3. Reporters for detecting nuclear receptor signaling can be classified into two main categories: (i) gene promoter–reporter fusions, and (ii) highly specific artificially constructed minimal promoters containing isolated hormone response elements.

   Gene specific promoter fusions (for instance use of the PSA or probasin promoter to study AR activity) [13–15] have the benefit of showing nuclear receptor activity in the "natural" context but are less useful to study receptor activity beyond the confines of the tissue where the gene is normally expressed (in this case the prostate), and often contain sites for multiple nuclear receptors and other potentially confounding transcription factors. Artificially constructed promoters with a minimal promoter region under control of inserted hormone response element(s) often have weak activity but have the benefits of being activated exclusively by the desired receptor so more likely to show global (whole organism) activity without tissue bias. Several hormone response elements should be tested in the final construct using in vitro transfections together with plasmids expressing single nuclear receptor constructs [16]. Ideally these should be tested in a variety of cell lines both completely lacking endogenous nuclear receptors and those with well documented expressed receptors (although care should be taken to ensure wild type receptors are expressed).

Additionally a variety of ligands should be used in isolation and in combination to ensure that no cross over activity occurs.

Nuclear receptors cannot activate gene transcription in complete isolation. Specific sequences are required for the association of cellular transcription machinery. A core or minimal promoter is an essential part of an artificially constructed promoter. These consist of one or more of three conservative sequences, i.e., TATA box, initiator region, GC-rich region, and binding site for RNA polymerase and should be located between −35 and +35 region with respect to the transcription start site. Core promoters should always be tested before the addition of a hormone response element, to exclude hidden sequences that may elicit endogenous activation, randomly created by ligation of multiple DNA fragments.

4. Where a transgene is being randomly integrated into the genome, insulators (or boundary elements) protect them from being influenced by the activity of the surrounding chromatin into which they integrate. Insulator sequences occur naturally in the genome and function as barriers to promoter–enhancer communication and can block the spread of repressive (or active) chromatin. Two well-characterized insulator sequences are the chicken β-globin hypersensitive site (HS4) and the matrix attachment region (MAR) [17]. "Natural" gene specific promoters may not require an insulator sequence and will show strong but tissue-specific gene induction. However, weaker artificial promoters will become silenced over time when integrated into the genome unless they are protected from positional effects.
5. When producing transgenic animals by integration, the transgene is linearized and purified from the prokaryotic vector sequence. 1–2 picoliter (pL) of DNA at a concentration of 1–2 ng/L is microinjected into the pronucleus of a fertilized mouse egg, afterwards the embryos are surgically transferred into the oviduct of pseudopregnant mice. For genotyping, DNA is typically isolated from mouse tail biopsies or ear notching and analyzed for the presence of the transgene by Southern blotting or PCR. Integration is usually at a random chromosomal location. Although easier to undertake "in house," less time consuming and less costly than targeted integration, the random integration procedure has been reported to produce acceptable first generation founders but with increasingly silenced transgene expression in subsequent generations.
6. For production of mice with the transgene targeted to a specific locus, the most widely used such locus is Rosa26, on chromosome 6. This has not yet been used in the generation of nuclear receptor reporter mice, and it should be considered that the high activity of this locus could potentially mask subtleties of expression. Our approach was to target to the Hprt

locus (on the X chromosome), which is less highly expressed. Embryonic stem (ES) cells are derived from 3.5-day-old mouse embryos, at the blastocyst stage of development. These cells are then disaggregated, and individual ES cells clones are grown. The DNA transgene is cloned into a targeting vector construct which is used to target the transgene to a specific chromosomal location. The transgene is integrated into the chromosome via homologous recombination. The ES cells may have an engineered loss-of-function mutation, which is repaired by the correct integration of the targeting construct along with its transgene. Alternatively the targeting vector may confer antibiotic resistance, e.g., neomycin. This process allows the selection of ES cells which are then microinjected into the blastocoele of 3.5-day-old embryos. The injected embryos are then transferred into the uterus of pseudopregnant females that will give birth to chimeric mice, which are partly derived from the ES cells and partly from the host embryo. Subsequent breeding steps from these chimeras are required to generate the strain carrying the transgene in all tissues.

7. The animal imaging procedure is essentially the same for nude mice harboring xenografts with integrated bioluminescent reporters as for transgenic mice with bioluminescent reporters (*see* Fig. 3). However, it should be noted that tumors will have different blood supply kinetics to normal tissue, and that xenografts with overexpressed or overactive nuclear receptors will show a heightened response compared with normal tissue. Therefore, the importance of establishing the plateau phase for maximal luciferase signal is of paramount importance.

8. If gonadectomy or adrenalectomy are required, surgical procedures should be carried out only by a trained expert or veterinarian. Animals must have sufficient time to recover from the procedure before any other procedures are carried out. This also gives sufficient time for endogenous hormone levels to become minimal, e.g., surgical castration of the testes with testosterone level follow-up. Castrated or gonadectomized mice may be purchased via Charles River or Harlan in the UK with a full permissive project license and surgical justification.

   Additional care should be given to the animals if the change in hormone levels results in decreased health or well-being, e.g., salt imbalance due to adrenalectomy, stress due to cortisol level decrease, and increased/decreased fighting in castrated males. Veterinary advice should be taken at all times and sufficient hormonal replacement available if required.

   Additionally, care should be taken with the choice of diet for the mice as some animal feeds are rich in plant materials and soya which may be a source of xenobiotics and phytoestrogens.

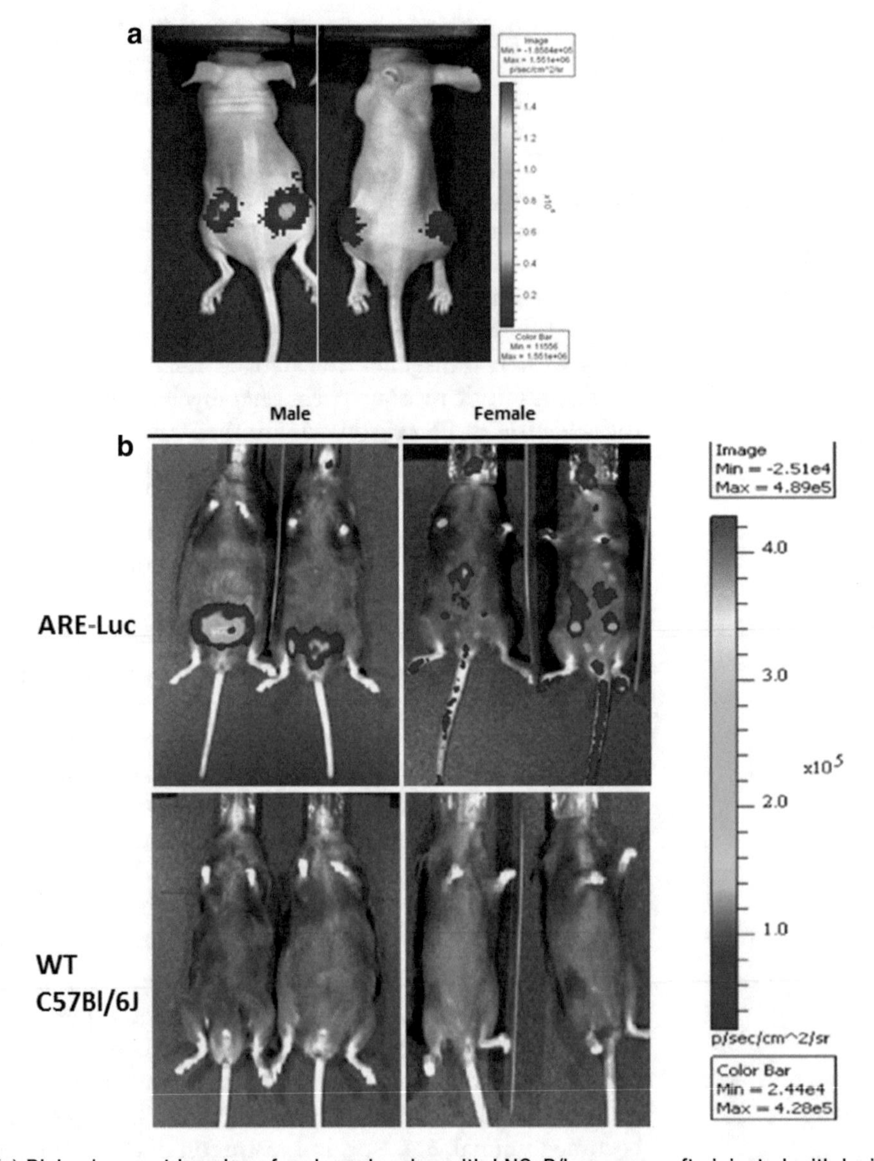

Fig. 3 (**a**) Bioluminescent imaging of nude male mice with LNCaP/Luc xenografts injected with luciferin substrate. Image represents a greyscale photograph overlaid with a pseudocolor image representing bioluminescent flux (photons/s/cm^2). (**b**) Luciferase imaging of male and female androgen receptor reporter transgenic mice (ARE-Luc). Image represents a greyscale photograph, overlaid with the photon emission intensity captured by CCCD camera, using a false color scale (photons/min/cm^2). Image reproduced in part from ref. 10

9. Different routes of hormone administration may be optimal dependent on solvent, pharmacokinetics, etc.:

   Daily injection of required hormone: A solution of hormone may be made in ethanol (or similar solvent) and then diluted for injection. Suitable substances are corn or sesame oil, and propylene glycol. The use of slowly metabolized hormonal precursors, e.g., testosterone propionate may be used to give a slow-release dose of the hormone. Hormone levels should be monitored by ELISA testing from collected serum.

Hormone Pellet: Pellets may be inserted under the skin by surgery under full anesthesia. However, it should be noted that once inserted they cannot easily be removed as they fragment over time. This is of importance in studies where hormone regimens may need to be changed during the experiment.

Single dosage example—for immediate induction of hormonal response (maximal at 6 h post injection with clearance at 24 h). This regimen is suitable for immediate hormonal response in tissues, but only for short term studies.

(i) Make 10 mM stock of hormone, e.g., estradiol (E_2) in 100% ethanol.

(ii) Dosage is 50 μg/kg and an average weight of a mouse is 25 g—therefore 1.25 μg/mouse.

(iii) 1.25 μg is equivalent to 0.005 μmol approx. (molecular weight 272.4 g/mol).

(iv) 10 mM stock = 0.01 μmol/ml and we need 0.005 μmol per mouse = 0.5 μl per mouse.

(v) Therefore, mix 0.5 μl of E_2 with 99.5 μl of oil for an injection volume of 100 μl per mouse.

Higher dosage example for slow-release hormone injection (maximal at 24–48 h and continuous).

The use of slowly metabolized hormonal precursors, e.g., testosterone propionate may be used to give a slow-release dose of the hormone. This procedure is more useful for xenograft or long term physiological studies. Note dosage is in μg/mouse units.

(i) Make a stock solution of hormone in ethanol as above (10 mM of testosterone propionate).

(ii) Dosage is 10–100 μg/mouse, every 24–48 h, depending on the serum level required.

(iii) For 10 μg/mouse = 10 μg is equivalent to 0.03 μmol approx.

(iv) 10 mM stock = 0.01 μmol/ml and we need 0.03 μmol per mouse = 3 μl per mouse.

(v) Therefore, mix 3 μl of E_2 with 97 μl of propylene glycol for an injection volume of 100 μl per mouse.

(vi) Or, for 50 μg/mouse, mix 15 μl of E_2 with 85 μl of propylene glycol for an injection volume of 100 μl per mouse.

10. When imaging cells, optimal luciferase activity will be close to instantaneous but sequential images may be taken to ensure this.

11. Most bioluminescent imaging equipment will only measure light photons which have hit the CCCD camera emanating from the surface of the subject—i.e., in two-dimensions only. The use of the IVIS Spectrum equipment and software enables the light emission to be analyzed at multiple wavelengths—enabling the tissue penetrance and subsequent depth of the signal to be calculated and reported in three dimensions with exact coordinates.

    The bioluminescent 3D image data may then be co-registered with anatomical data or images to reveal the source of the bioluminescent signal. Advanced tomographic imaging allows the bioluminescent signal to be co-registered with either an available digital mouse atlas—a generic mouse physiology model, or can indeed be co-registered with imported computerized tomography (CT) X-ray images of the actual subject animal (or indeed magnetic resonance imaging of the same animal).

References

1. Makkonen H, Jaaskelainen T, Rytinki MM, Palvimo JJ (2011) Analysis of androgen receptor activity by reporter gene assays. Methods Mol Biol 776:71–80. doi:10.1007/978-1-61779-243-4_5
2. Rossant J, Zirngibl R, Cado D, Shago M, Giguere V (1991) Expression of a retinoic acid response element-hsplacZ transgene defines specific domains of transcriptional activity during mouse embryogenesis. Genes Dev 5(8):1333–1344
3. Montoliu L, Blendy JA, Cole TJ, Schutz G (1995) Analysis of perinatal gene expression: hormone response elements mediate activation of a lacZ reporter gene in liver of transgenic mice. Proc Natl Acad Sci U S A 92(10): 4244–4248
4. Ciana P, Di Luccio G, Belcredito S, Pollio G, Vegeto E, Tatangelo L, Tiveron C, Maggi A (2001) Engineering of a mouse for the in vivo profiling of estrogen receptor activity. Mol Endocrinol 15(7):1104–1113
5. Klotz DM, Hewitt SC, Ciana P, Raviscioni M, Lindzey JK, Foley J, Maggi A, DiAugustine RP, Korach KS (2002) Requirement of estrogen receptor-alpha in insulin-like growth factor-1 (IGF-1)-induced uterine responses and in vivo evidence for IGF-1/estrogen receptor cross-talk. J Biol Chem 277(10):8531–8537. doi:10.1074/jbc.M109592200
6. Ciana P, Raviscioni M, Mussi P, Vegeto E, Que I, Parker MG, Lowik C, Maggi A (2003) In vivo imaging of transcriptionally active estrogen receptors. Nat Med 9(1):82–86
7. Ciana P, Biserni A, Tatangelo L, Tiveron C, Sciarroni AF, Ottobrini L, Maggi A (2007) A novel peroxisome proliferator-activated receptor responsive element-luciferase reporter mouse reveals gender specificity of peroxisome proliferator-activated receptor activity in liver. Mol Endocrinol 21(2):388–400. doi:10.1210/me.2006-0152
8. Houten SM, Volle DH, Cummins CL, Mangelsdorf DJ, Auwerx J (2007) In vivo imaging of farnesoid X receptor activity reveals the ileum as the primary bile acid signaling tissue. Mol Endocrinol 21(6):1312–1323. doi:10.1210/me.2007-0113
9. Ye X, Han SJ, Tsai SY, DeMayo FJ, Xu J, Tsai MJ, O'Malley BW (2005) Roles of steroid receptor coactivator (SRC)-1 and transcriptional intermediary factor (TIF) 2 in androgen receptor activity in mice. Proc Natl Acad Sci U S A 102(27):9487–9492. doi:10.1073/pnas.0503577102
10. Dart DA, Waxman J, Aboagye EO, Bevan CL (2013) Visualising androgen receptor activity in male and female mice. PLoS One 8(8):e71694. doi:10.1371/journal.pone.0071694
11. Contag CH, Spilman SD, Contag PR, Oshiro M, Eames B, Dennery P, Stevenson DK, Benaron DA (1997) Visualizing gene expression in living mammals using a bioluminescent reporter. Photochem Photobiol 66(4):523–531

12. Workman P, Aboagye EO, Balkwill F, Balmain A, Bruder G, Chaplin DJ, Double JA, Everitt J, Farningham DA, Glennie MJ, Kelland LR, Robinson V, Stratford IJ, Tozer GM, Watson S, Wedge SR, Eccles SA (2010) Guidelines for the welfare and use of animals in cancer research. Br J Cancer 102(11):1555–1577. doi:10.1038/sj.bjc.6605642
13. Hsieh CL, Xie Z, Liu ZY, Green JE, Martin WD, Datta MW, Yeung F, Pan D, Chung LW (2005) A luciferase transgenic mouse model: visualization of prostate development and its androgen responsiveness in live animals. J Mol Endocrinol 35(2):293–304. doi:10.1677/jme.1.01722
14. Seethammagari MR, Xie X, Greenberg NM, Spencer DM (2006) EZC-prostate models offer high sensitivity and specificity for noninvasive imaging of prostate cancer progression and androgen receptor action. Cancer Res 66(12):6199–6209. doi:10.1158/0008-5472.CAN-05-3954
15. Ellwood-Yen K, Wongvipat J, Sawyers C (2006) Transgenic mouse model for rapid pharmacodynamic evaluation of antiandrogens. Cancer Res 66(21):10513–10516. doi:10.1158/0008-5472.CAN-06-1397
16. Dart DA, Spencer-Dene B, Gamble SC, Waxman J, Bevan CL (2009) Manipulating prohibitin levels provides evidence for an in vivo role in androgen regulation of prostate tumours. Endocr Relat Cancer 16(4):1157–1169. doi:10.1677/ERC-09-0028
17. Bell AC, West AG, Felsenfeld G (2001) Insulators and boundaries: versatile regulatory elements in the eukaryotic genome. Science 291(5503):447–450

Chapter 14

Development and Characterization of Cell-Specific Androgen Receptor Knockout Mice

Laura O'Hara and Lee B. Smith

Abstract

Conditional gene targeting has revolutionized molecular genetic analysis of nuclear receptor proteins, however development and analysis of such conditional knockouts is far from simple, with many caveats and pitfalls waiting to snare the novice or unprepared. In this chapter, we describe our experience of generating and analyzing mouse models with conditional ablation of the androgen receptor (AR) from tissues of the reproductive system and other organs. The guidance, suggestions, and protocols outlined in the chapter provide the key starting point for analyses of conditional-ARKO mice, completing them as described provides an excellent framework for further focussed project-specific analyses, and applies equally well to analysis of reproductive tissues from any mouse model generated through conditional gene targeting.

Key words Cre/lox, Transgenic, Mouse, Conditional knockout, Androgen receptor, ARKO

1 Introduction

1.1 Development of Transgenic Technologies and ES Cell Knockouts

Over the past three decades, genetic manipulation in the mouse has transformed our understanding of sex-hormone nuclear receptor function [1, 2]. Knockouts, knockins, gene-trapped, ENU-derived and other transgenic mouse models have allowed us to delicately dissect many aspects of the roles nuclear receptor signaling (and related genes) plays in reproductive development and function (for reviews *see* refs. 3–7).

1.2 Development of the Cre/Lox System

Due to the widespread and all-pervading nature of nuclear receptor signaling, the development of the Cre/*Lox* system of conditional gene targeting over 20 years ago [8, 9], has been instrumental in dissecting out and attributing specific functions (for reviews *see* refs. 10, 11). The ability of this system to temporally or spatially restrict genetic manipulation has provided a step-change over previous knockout models. Whilst other recombination systems are available [12], Cre has become the recombinase of choice primarily because it recombines DNA at high efficiency, and does not require

Iain J. McEwan (ed.), *The Nuclear Receptor Superfamily: Methods and Protocols*, Methods in Molecular Biology, vol. 1443, DOI 10.1007/978-1-4939-3724-0_14, © Springer Science+Business Media New York 2016

specific DNA topology or accessory proteins to mediate efficient recombination in target molecules [13]. The utility of this system has been significantly enhanced by the critical mass of usage that has developed over the past 10 years. There are now greater than 500 independent Cre Recombinase expressing mouse lines, many of which are freely available via the CreXmice database [14].

The concept behind the *Cre/Lox* system is based upon Cre Recombinase's ability to bind and recombine DNA between two *LoxP* (Locus Of crossing [X-ing] over in P1) sites, each of these 34 bp target DNA sequences is made up of two 13 bp inverted repeat sequences, flanking a central, 8 bp, directional core. Use of this technology in mammalian systems is possible because the mammalian genome does not possess high affinity *LoxP* sites (low-affinity *LoxP* sites do occur naturally in the mammalian genome, but are recombined at extremely low efficiency if at all [15, 16]), therefore, engineering *LoxP* sequences into DNA regions flanking a target gene or exon "highlights" that site as a target for recombination by Cre Recombinase [17] (Fig. 1).

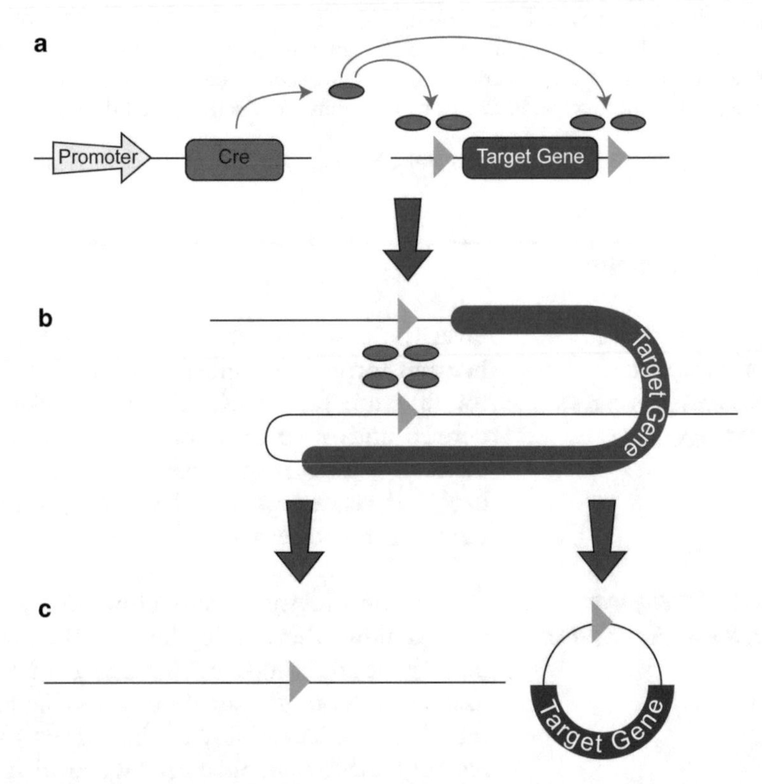

Fig. 1 The mechanism of DNA recombination via the Cre/loxP system (**a**) Cre recombinase (driven by an exogenous promoter of choice) enters the cell nucleus and binds to LoxP sites flanking a target gene. (**b**) The DNA surrounding the target gene undergoes a conformational change to bring the two LoxP sites into close proximity. A synaptonemal complex is formed where DNA strand exchange occurs at the sites of homology leading to the excision of the target gene (**c**) A single LoxP site (generated from two half-sites) is left at the original genomic location

Table 1
High-throughput development projects producing commercially available Cre-loxP lines

Name	Website
International Mouse Phenotyping Consortium (IMPC)	https://www.mousephenotype.org
Knockout Mouse Project (KOMP)	https://www.komp.org/
North American Conditional Mouse Mutagenesis Project (NorCOMM)	http://www.norcomm.org/
European Conditional Mouse Mutagenesis Program (EUCOMM)	http://www.mousephenotype.org/martsearch_ikmc_project/about/eucomm
European Mouse Disease Clinic (EUMODIC)	http://www.eumodic.org/
European Mouse Mutant ARchive (EMMA)	https://www.infrafrontier.eu
The Jackson Laboratory (JAX)	http://www.jax.org/
Mutant Mouse Regional Resource Centers (MMRRC)	https://www.mmrrc.org/

Using this nomenclature, target genes flanked by *LoxP* sites are said to be "*floxed*" [12, 18, 19]. As with the explosion in Cre Recombinase-expressing mouse lines, the utility and take-up of this system has been significantly enhanced by the ongoing development of floxed alleles for all protein coding genes in the mouse genome. These high-throughput development pipelines are largely public or charity funded, with resulting floxed ES cells made freely available to anyone (Table 1).

Together, the widespread availability of both Cre and floxed mice means it is now relatively simple for a researcher to obtain both a floxed gene of interest, and a Cre line to permit targeting of that gene in any given lineage of the mouse. Like PCR, conditional gene targeting is now seen as "just another tool" of the molecular geneticist's toolbox. Though as one might expect, idiosyncrasies of the technology in practice make use of this system a minefield for the uninitiated (for review *see* ref. 20). These caveats are discussed in detail below, using our generation and characterization of conditional gene targeting of the Androgen Receptor (AR) across several reproductive cells and tissues.

1.3 Choice of Cre Lines

For recombination to take place in vivo, Cre Recombinase itself must be expressed within the target cells. Several approaches have been developed to ensure this is the case. The most widespread is the generation of transgenic mouse lines via pronuclear injection [21] containing the Cre Recombinase gene downstream of a promoter fragment from a gene with established expression in the target cells/tissues. It is this choice of promoter that primarily defines specificity of Cre expression. Generation of transgenics via pronuclear injection is a well-established and speedy method of

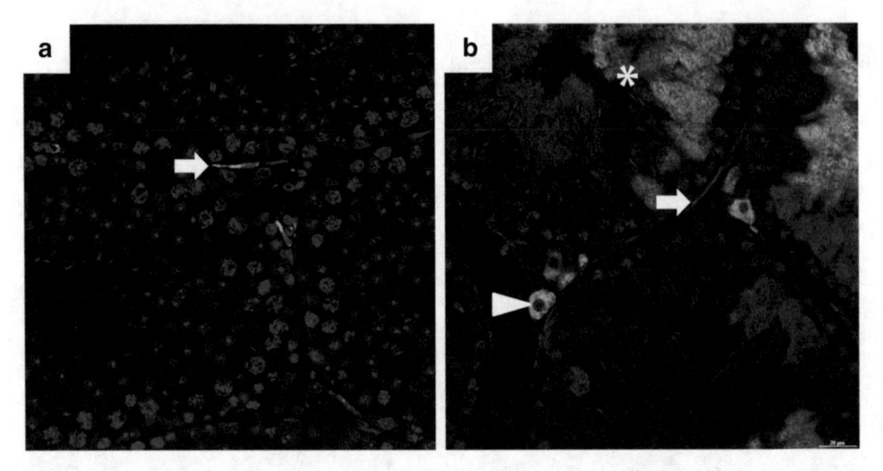

Fig. 2 Localization of Cre recombination in a tissue may be different from expression of the endogenous promoter (**a**) Immunolocalization of nestin (*in green*) in a wild-type adult testis section is in a subpopulation of peritubular cells (*arrow*). (**b**) Immunolocalization of YFP (*in green*) in a Nestin-Cre x R26RYFP testis section is in a subpopulation of peritubular cells (*arrow*) but also in a subpopulation of Leydig cells (*arrowhead*) and germ cells (*asterisk*). Nuclear counterstain is in *blue*

producing transgenic animals, and as such is very attractive in this respect. However generation of Cre expressing lines in this way has several disadvantages. In the majority of Cre lines, for technical reasons promoter fragments are used, and as such, actual expression of Cre Recombinase can be significantly different from the endogenous gene. In addition, such transgenes insert randomly in the genome, which can lead to transgene silencing, either due to insertion in a region of the genome heterochromatic at the time Cre expression is required, or via epigenetic modification, often as transgenes are passed on to the next generation. In addition, transgenes rarely insert in single copies, but favor arrays of inserted transgenes at the same genomic locus, which can complicate downstream analysis or affect cell viability. Very few Cre lines perfectly recapitulate the endogenous expression pattern (Fig. 2). Presuming that they will work as designed has been a major issue in terms of the success of *Cre/Lox* experiments.

In contrast to this, careful empirical validation of Cre Recombinase sites of expression (even if ectopic) can provide serendipitous benefits. For example, through inclusion of appropriate controls we have been able to utilize the same smMHC-Cre Recombinase line to determine the role of AR-signaling in separate studies looking at the peritubular myoid cells [22, 23], seminal vesicle smooth muscle [24] and prostate smooth muscle [25].

In some circumstances, researchers have instead "knocked-in" Cre to a specific locus in embryonic stem (ES) cells, using the endogenous promoter to try and more faithfully follow the endogenous expression pattern [21]. A caveat of this approach is that Cre Recombinase is inserted as a single copy gene, and as such is then reliant on the strength (and timing) of the promoter to ensure sufficient gene expression to permit DNA recombination. Other considerations when deciding whether a knocked-in Cre line is a good choice is that this approach is often utilized to simultaneously generate a null allele of the gene Cre Recombinase is knocked-in to (and, through breeding of two mice carrying the transgene, to generate a knockout of the endogenous gene when homozygous), in addition to generating a useful Cre Recombinase expressing line when heterozygous [14]. This can lead to haploinsufficiency of the endogenous gene in Cre Recombinase carrying mice, which can prove a confounding factor in downstream analysis. To avoid this, several researchers have instead inserted Cre Recombinase downstream of an Internal Ribosome Entry site (IRES) place in the 3′ UTR of the endogenous gene. These "gene-trap" lines can provide reliable promoter-driven expression without ablation of the endogenous gene [14], though use of an IRES often results in reduced expression of the downstream gene, which may affect the concentration of Cre Recombinase produced in a cell.

1.4 Validation of Cre Expression and Function

As mentioned above, expression of Cre Recombinase in many lines does not follow the expected expression pattern expected of the endogenous gene from which their promoter has been taken. Thus, Cre lines should always be considered in terms of their utility rather than in terms of their original purpose. To establish whether Cre lines will be of use to a researcher, the key point that cannot be overstated is the need for empirical validation of expression. Validating the spatial and temporal expression of Cre Recombinase in a mouse line can be accomplished by the combination of several methodological approaches including immunohistochemistry against Cre Recombinase, or if a fluorescent marker is included [14], by immunohistochemistry against this marker. Whilst direct examination of fluorescence itself is a good indicator of expression, this rarely provides the level of resolution necessary to attribute expression to individual cell-types (see below). The caveat of this approach is that this only establishes where the Cre Recombinase is *currently* expressed, not where is has been expressed and functioned in the past. Lineage-tracing using a Cre-inducible reporter line (of which there are many, including lacZ [26–32], fluorescent reporter proteins [33–41], and more recently bioluminescent proteins [42–44]) permits persistent tracking of daughter cells arising from any cell in which a recombination event has taken place (Fig. 3). This is often a better indicator of the utility of the Cre line, though this itself is not without limitations. Many reporter lines utilize the easy accessibility of the Rosa26 locus, and we have found that expression of reporter genes from this permissive locus can be activated by Cre Recombinase more readily than floxed genes of interest located in other regions of the genome [21] (Fig. 4). A second issue to bear in mind is that increased reporter expression in one cell-type over another does not mean

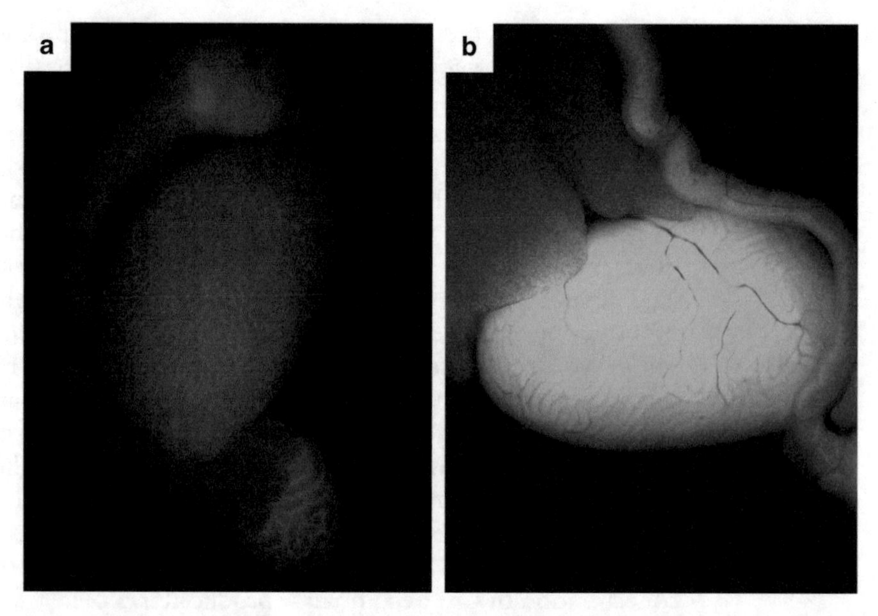

Fig. 3 Examples of fluorescent protein expression in the testis. Testes from two mouse lines, (**a**) one expressing tdTomato fluorescent protein and (**b**) one expressing yellow fluorescent protein both viewed with a microscope fitted with an epifluorescent filter specific for the protein

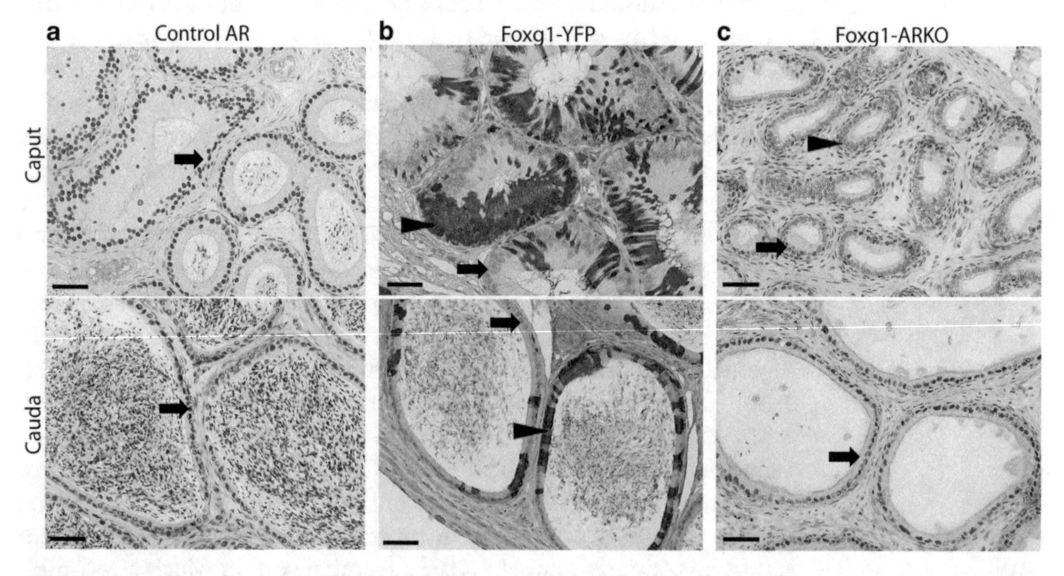

Fig. 4 Localization of Cre recombination in a tissue may differ between loci (**a**) Androgen receptor (*brown*) is expressed throughout the epithelium of the caput and cauda epididymis (*arrows*). (**b**) Immunolocalization of YFP (*brown*) in a Foxg1-Cre x R26RYFP testis section is in a subpopulation of epididymal cells in both the caput and cauda epididymis (*arrowheads*) but not present in others (*arrows*). (**c**) In Foxg1-Cre x AR^{flox} epididymides, AR is ablated in a subpopulation of caput cells (*arrowhead*) but is still seen others (*arrow*), but no ablation of AR is seen in the epithelium of the cauda epididymis (*arrow*). Scale bars are 50 μm

"better" targeting in that cell-type. Differences in "level" of expression of the reporter may simply reflect the cellular turnover of the reporter protein in different cell-types. The Cre/lox system is largely binary, you either have recombination or you do not.

1.5 Cre Recombinase Can Itself Produce a Phenotype

Once the spatiotemporal pattern of Cre expression has been confirmed, and deemed to have utility, the first control in any study is to determine whether the Cre Recombinase line itself induces a phenotype in the mouse. Such an effect has been demonstrated most notably in use of the RIP-Cre mouse [45]. We examine this by aging a cohort of Cre Recombinase expressing mice beyond the endpoint of the planned experiment, and demonstrating that, in the absence of LoxP sites, inheritance of this Cre Recombinase transgene has no phenotypic impact. An alternative to specifically aging a separate cohort is to use Cre Recombinase stud males which are often kept to breed to floxed females throughout the extent of the study and can be used once the experiment is complete. This can prove cost saving if no phenotype is present, or disastrous if the Cre Recombinase alone results in confounding impacts on the phenotype. Individual choice and risk–benefit analysis of taking this approach should be considered carefully.

Described below is our standard methodological approach to generation and characterization of cell-specific ARKO models. This covers the empirical validation of Cre Recombinase mouse lines, and the first-pass phenotypic analysis of resultant cell-specific ARKO mice. The majority of studies can be separated into three simple parts: (1) Generation and validation of the model; (2) characterization of the resultant phenotype; and (3) results-led, and project-specific functional analysis and rescue of the phenotype. The combination of these three project strands provides confidence that the project has been completed to a high standard.

2 Materials

2.1 Validation of Cre Recombinase Expression Through Activation of a Reporter Gene

2.1.1 Breeding of Fluorescent Reporter Mice

1. R26R-EYFP and cell specific Cre recombinase mouse lines are available from the suppliers detailed in Table 1.

2.1.2 Viewing Fluorescent Reporter Expression in Whole Tissues Under an Epifluorescent Microscope

1. Phosphate buffered saline ("PBS"); One PBS tablet (Sigma-Aldrich #P4417) dissolved in 200 mL of deionized water yields 0.01 M phosphate buffer, 0.0027 M potassium chloride, and 0.137 M sodium chloride, pH 7.4, at 25 °C.
2. Leica MZFLIII microscope with YFP filter system.

2.1.3 Localizing Fluorescent Reporter Expression by Immunohistochemistry

1. Bouin's fixative (Clin-Tech, Guildford, UK).
2. Absolute ethanol.
3. Rotary microtome (Leica Biosystems).
4. Water bath.
5. Small artist's paintbrushes.
6. Coated glass slides (A-pex, Leica Biosystems).
7. Slide oven.
8. Xylene.
9. Citrate buffer; 10 mM sodium citrate, adjusted to pH 6.
10. Decloaking Chamber (Biocare Medical).
11. Tris-buffered saline ("TBS"); 60.5 g Trizma base, 87.6 g sodium chloride, 300 mL hydrochloric acid, made up to 10 L with distilled water, adjusted to pH 7.4. This makes a 10× solution that should be diluted 1 in 10 before use.
12. NGS/TBS/BSA; 2 mL normal goat serum ("NGS," Biosera), 8 mL TBS, 0.5 g bovine serum albumin ("BSA," Sigma).
13. Rabbit anti-GFP/YFP antibody (Abcam #ab6556).
14. TBS-T; TBS, 0.025% Triton X-100.
15. 30% hydrogen peroxide.
16. Goat anti-rabbit peroxidase (DAKO #P0448).
17. Cyanine 3-conjugated Tyramide ("TSA," PerkinElmer).
18. SYTOX Green (#P4170, Sigma, UK).
19. PermaFluor (Thermo Scientific).
20. Glass coverslips (Thermo Scientific).
21. LSM 710 confocal microscope (Zeiss) with Zen software.

2.2 Breeding and Genotyping of Conditional ARKO Mice

2.2.1 Breeding of Conditional ARKO Mice

1. AR^{flox} and cell-specific Cre recombinase mouse lines are available from the suppliers listed in Table 1.

2.2.2 Determination of Inheritance of Cre Transgene by PCR

1. TE Tween buffer; 2.5 mL 1 M Tris (pH 8.0), 100 μL 0.5 M EDTA (pH 8.0), 250 μL Tween 20, adjusted to 50 mL with distilled water.
2. Proteinase K (Roche, #03115879001 10 mg/mL).
3. Sterile water (Sterac).
4. Primers detailed in Table 2 (MWG Operon).
5. BioMix Red (Bioline).

Table 2
PCR Genotyping assays used in breeding of Cre and floxed lines

Assay name	Primers	Band sizes expected
R26R-EYFP To distinguish between homozygote, heterozygote, and wild-type Rosa26 insertion	F: GGAGCGGGAGAAATGGATATG R1: AAAGTCGCTCTGAGTTGTTAT R2: AAGACCGCGAAGAGTTTGTC	WT Rosa26: 600 bp Rosa26 insertion: 320 bp
AR^{flox} To distinguish between homozygote, heterozygote, wild-type, and recombined AR^{flox}	F: GCTGATCATAGGCCTCTCTC R: TGCCCTGAAAGCAGTCCTCT	Floxed AR: 1142 bp, WT AR: 1072 bp Recombined AR: 612 bp
Cre genotyping To determine whether Cre transgene has been inherited	F: GCGGTCTGGCAGTAAAAACTATC R: GTGAAACAGCATTGCTGTCACTT	102 bp amplicon when Cre is present

6. Thermocycler.
7. QIAxcel capillary genotyping machine (Qiagen).

2.3 Phenotypic Examination of Conditional ARKO Mice

2.3.1 Indications of Disrupted Androgen Receptor Action During Gross Dissection

1. Dissection tools.
2. Electronic digital caliper (Faithfull Tools).
3. Balance sensitive to 0.001 g (Sartorius).
4. Bouin's fixative (Clin-Tech, Guildford, UK).
5. Dry ice.

2.3.2 Determination of Genomic Ablation of AR in Tissues of Interest by PCR

1. Column-based genomic DNA extraction kit (Thermo Scientific).
2. Thermocycler.
3. Primers detailed in Table 2 (MWG operon).
4. BioMix Red (Bioline).
5. QIAxcel capillary genotyping machine (Qiagen).

2.3.3 Determination of Spatial Ablation of AR by Immunohistochemistry

1. For materials *see* Subheading 2.1.3.
2. Rabbit anti-AR antibody (#M4070, Spring Bioscience).

2.3.4 Histological Analysis of Conditional ARKO Tissues

1. Xylene.
2. Absolute ethanol.
3. Harris' hematoxylin (Leica Biosystems).
4. Acid–alcohol; 1% concentrated hydrochloric acid in 70% ethanol.

5. Scott's tap water; 0.2% w/v potassium hydrogen carbonate, 2% w/v magnesium sulfate, tap water.
6. Aqueous Eosin (Leica Biosystems).
7. Pertex (Leica Biosystems).
8. Glass coverslips (Thermo Scientific).
9. Light microscope.

2.3.5 Quantification of Testicular Cell Types

1. Olympus BX50 microscope (Prior Scientific Instruments, Cambridge, UK).
2. Stereologer software (Systems Planning Analysis, Alexandria, VA, USA).

2.3.6 Analysis of Circulating Hormone Levels

1. 25G × 5/8 in. gauge needles (Becton Dickinson).
2. 1 mL syringes (Becton Dickinson).
3. Heparin (LEO Pharma).
4. Small glass test tubes (to fit your own centrifuge).
5. CHROMASOLV diethyl ether (Sigma-Aldrich 309966-100ML).
6. Testosterone ELISA kit (DEV9911, Demeditec Diagnostics, Germany).
7. FSH ELISA kit (Cusabio Life Science, CSB-E06871m).
8. LH ELISA kit (Cusabio Life Science, CSB-E12770m).
9. Microplate spectrophotometer (BioTek).
10. ELISA Analysis software (www.elisaanalysis.com).

2.3.7 Fertility Testing

1. No specific materials required.

2.3.8 Assessment of Leydig Cell Maturation Using qRT-PCR

1. RNeasy Mini extraction kit (Qiagen).
2. RNase-free DNase on the column digestion kit (Qiagen).
3. Luciferase mRNA (Promega #L4561).
4. NanoDrop 1000 spectrophotometer (Thermo Fisher Scientific).
5. SuperScript VILO cDNA synthesis kit (Life Technologies).
6. Taqman® Gene Expression Mastermix (Life Technologies #4369016).
7. Mouse Universal Probe Library (Roche).
8. Primers for assay of interest as detailed in Table 4 (MWG Operon).
9. ABI Prism 7900 (Life Technologies).
10. ABI Sequence Detection System software (Life Technologies).

3 Methods

3.1 Validation of Cre Recombinase Expression Through Activation of a Reporter Gene

3.1.1 Breeding of Fluorescent Reporter Mice

1. Maintain R26R-EYFP reporter mice [39] in an inbred homozygous colony
2. Verify homozygosity every 3 months by genotyping from ear clips as detailed in Subheading 3.2.2 using the PCR primer assay described in Table 2 to distinguish between homozygote, heterozygotes, and wild types. This protects against accidental introduction of non-homozygous individuals to the colony.
3. Mate stud males heterozygous for the Cre Recombinase transgene of interest with R26R-EYFP homozygous females.
4. All offspring are heterozygous for the R26R-EYFP transgene. Half of these offspring are also heterozygous for the Cre Recombinase transgene (and thus suitable for lineage tracing) and half do not carry Cre Recombinase (used as controls to determine sites/intensity of autofluorescence). Since one copy of the R26R-EYFP transgene per cell is sufficient for lineage tracing, mice do not need to be bred for further generations.

3.1.2 Viewing Fluorescent Reporter Expression in Whole Tissues Under an Epifluorescent Microscope

1. Postmortem, remove tissues from the offspring of matings between Cre and R26R-EYFP mice to phosphate buffered saline (PBS) on ice (*see* **Note 1**).
2. These tissues can be viewed under a dissecting microscope fitted with an epifluorescent filter system, such as the Leica MZFLIII (*see* Fig. 3).
3. View and image the same tissues from a Cre-positive littermate and a Cre-negative littermate side-by-side (*see* **Note 2**) to avoid mistaken identification of background tissue autofluorescence as positive fluorescence (Fig. 5).

3.1.3 Localizing Fluorescent Reporter Expression by Immunohistochemistry

Immunohistochemistry is performed to localize fluorescent reporter to specific cell types in tissues that appeared YFP-positive when viewed under epifluorescence.

1. Fix tissues for 6 h in Bouin's fixative (*see* **Note 3**).
2. Remove tissues to 70% ethanol (*see* **Note 4**).
3. Embed tissues in paraffin and mount in microtome blocks (*see* **Note 5**).
4. Cut 5 μm thick sections in ribbons from paraffin blocks using a rotary microtome.
5. Float ribbons on the surface of a water bath set at 45 °C. Separate individual sections in the ribbon by pulling apart using small paintbrushes and then transfer sections to charged glass slides by dipping the slide into the water under-

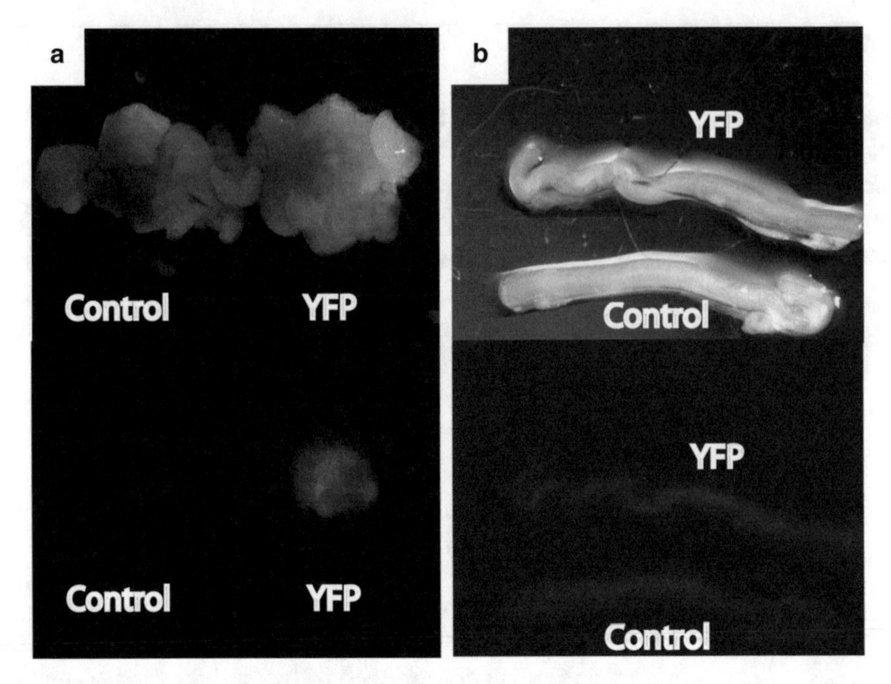

Fig. 5 Tissue autofluorescence (**a**) when viewed with a microscope fitted with an epifluorescent filter for YFP, fluorescence can clearly be seen in Cyp11a1-Cre x R26RYFP ovaries, but not in control ovaries. (**b**) Autofluorescence of the vas deferens can be seen in both controls and Cyp11a1-Cre x R26RYFP mice (which should not express YFP in the vas deferens)

neath the section and gently lifting until contact is made between the slide and the section.

6. Incubate slides in an oven set at 55 °C overnight to allow sections to dry and adhere to slides.
7. Dewax slides by incubating in a xylene bath for 5 min (*see* **Note 6**).
8. Move slides to a new xylene bath for 5 min.
9. Move slides to a 90% ethanol bath for 20 s.
10. Move slides to a 70% ethanol bath for 20 s.
11. Wash slides in tap water for 20 s.
12. Perform heat induced epitope retrieval in citrate buffer (*see* **Note** 7) in a pressurized Decloaking Chamber (*see* **Note 8**).
13. Cool slides in tap water.
14. Block nonspecific IgG binding sites by covering sections with NGS/TBS/BSA (*see* **Note 9**) and incubating for 30 min minimum in a humidified slide chamber at room temperature.
15. Gently wipe the serum from around the sections and then cover the sections with rabbit anti-YFP antibody diluted to a predetermined titered concentration (*see* **Note 10**) in NGS/TBS/BSA. Incubate slides overnight at 4 °C in the humidified slide chamber.

16. Wash slides for 5 min in TBS-T for 5 min, then twice in TBS.
17. Incubate slides in TBS with 3% v/v hydrogen peroxide for 15 min on a rocking platform to saturate any endogenous peroxidase.
18. Rinse slides with water, then cover sections with goat anti-rabbit peroxidase-tagged antibody diluted to a predetermined concentration in NGS/TBS/BSA. Incubate slides for 30 min minimum in humidity chamber at room temperature.
19. Wash slides for 5 min in TBS-T for 5 min, then twice in TBS.
20. Cover sections with Cyanine 3-conjugated Tyramide, diluted 1:50 in the included kit substrate and incubate slides for 10 min to visualize antibody staining in red (*see* **Note 11**).
21. Wash slides in TBS three times for 5 min each.
22. Cover sections with SYTOX Green diluted 1 in 1000 in TBS and incubate for 10 min to stain nuclei green (*see* **Note 12**).
23. Wash slides in TBS three times for 5 min, then mount with PermaFluor and a glass coverslip.
24. Capture images using a Zeiss LSM 710 confocal microscope with Zen software.

3.2 Breeding and Genotyping of Conditional ARKO Mice

At least five "floxed" alleles of AR have been generated [46–50], (reviewed in refs. 51, 52). Together, these five AR alleles have been widely used to address the cell-specific roles of androgen-signaling in the testis and other reproductive tissues (reviewed in refs. 51–56). Two of these floxed AR models have loxP sites flanking exon 2 which encodes the DNA-binding domain [46, 50]. In the absence of recombination, these floxed male mice are identical to non-transgenic males. In contrast, Cre-mediated recombination leading to excision of exon 2 results in a frame shift mutation, introducing a premature stop codon into the mRNA transcript, which is rapidly degraded; no AR protein is produced. We have utilized the model generated by De Gendt and Verhoeven [46]. Generation of this mouse line carrying a floxed AR has been described previously [57]. The structure of the allele can be seen in Fig. 6.

To generate conditional ARKO mice, breed AR^{flox} mice with validated Cre Recombinase mouse lines to target various cells and tissues of interest (Table 3). Mouse lines containing Cre driven by various promoters and mouse lines containing floxed alleles of interest are available from the suppliers detailed in Table 1.

3.2.1 Breeding of Conditional ARKO Mice

1. Maintain AR^{flox} mice [46] in an inbred homozygous colony.
2. Verify homozygosity every 3 months by genotyping from ear clips as detailed in Subheading 3.2.2 using the PCR primer assay described in Table 2 to distinguish between homozygote, heterozygotes, and wild types. This protects against accidental introduction of non-homozygous individuals to the colony.

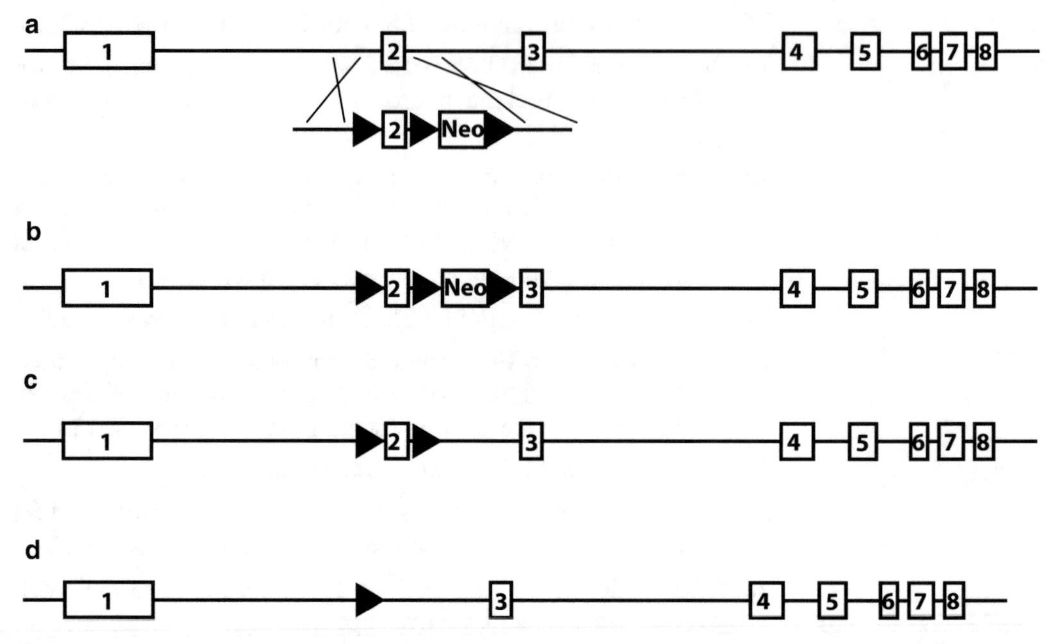

Fig. 6 Creation of the AR^{flox} allele (**a**) Wild type murine AR consists of eight exons (*numbered*). (**b**) A transgenic androgen receptor with loxP sites flanking exon 2 is created by homologous recombination in ES cells of a transgene consisting of loxP sites flanking both AR exon 2 and a downstream neomycin resistance gene. (**c**) The AR^{flox} allele is created when transient expression of Cre removes the neomycin resistance gene in ES cells. (**d**) In offspring carrying the AR^{flox} allele, recombination of the loxP sites flanking exon 2 results in creation of a null allele

Table 3
Cell/tissue targets of AR ablation and Cre Recombinase line used

Cell/tissue targeted	Cre-line used	ARKO Reference
Sertoli cell	Amh-Cre	[46, 47, 58]
Sertoli cell	Abp-Cre	[59]
Peritubular myoid cell	smMHC-Cre	[22]
Arteriole smooth muscle cells	SM22-Cre	[59]
Vascular endothelial cells	Tie2-Cre	[60]
Leydig cell (Sertoli cell?)	AmhR2-Cre	[61, 62]
Germ cell	Sycp1-Cre	[61]
Prostate smooth muscle	smMHC-Cre	[25]
Prostate epithelium	Probasin-Cre	[63]
Seminal vesicle smooth muscle	smMHC-Cre	[24]
Seminal vesicle smooth muscle	Probasin-Cre	[63]
Epididymis epithelium	Foxg1-Cre	[64]

(continued)

Table 3 (continued)

Cell/tissue targeted	Cre-line used	ARKO Reference
Epididymis epithelium	Rnase10-Cre	[65]
Gubernaculum	Rarb-Cre	[66]
Adipose tissue	Fabp4-Cre	[67, 68]
Adipose tissue	Adipoq-Cre	[67]
B lymphocytes	Cd19-Cre	[69]
T lymphocytes	Lck-Cre	[70]
Monocytes and granulocytes	Lyz-Cre	[71]
Keratinocytes (epithelial cells)	Ck5-Cre	[71]
Fibroblasts	Fsp1-Cre	[71]
Thymic epithelial cells	Ck5-Cre	[70]
Hepatocytes	Albumin-Cre	[72]
Neurons	SynapsinI-Cre	[73]
Neurons	Nestin-Cre	[74]
Osteoblasts	Col2.3-Cre	[75]
Osteoblasts	Osteocalcin-Cre	[76]
Odontoblasts, osteocytes, and muscle	Dmp1-Cre	[77]
Satellite cells	MyoD-Cre	[78]
Myocytes	Mck-Cre	[79]
Skeletal muscle cells	Hsa-Cre	[80]
Total ARKO	PGK1-Cre	[46]
Total ARKO	HPRT-Cre	[81]
Total ARKO	CMV-Cre	[82]
Total ARKO	Actb-Cre	[61]
Adult ARKO (tamoxifen-induced)	CreERT2	[83]

3. Mate stud males heterozygous for the Cre Recombinase transgene of interest with females homozygous for the AR^{flox} transgene.
4. As AR is X-linked, all male offspring are AR^{flox}/y hemizygotes (*see* **Note 13**). Half of the offspring are Cre-positive (and thus conditional knockouts for AR in the cell type of interest) and half Cre-negative.

5. As with the Cre Recombinase lines, AR^{flox} mice have been aged and extensively examined for phenotypic changes related to introduction of the floxed allele. As no phenotype has been observed, these littermates are used as controls.

3.2.2 Determination of Inheritance of Cre Transgene by PCR

A basic DNA extraction can be performed on the small piece of tissue obtained from an ear clip and genotyping performed to identify if the Cre transgene has been inherited.

1. At weaning at 21 days post birth, ear-clip offspring from matings between Cre and floxed mice for identification purposes.
2. Add the ear clip to a tube containing 25 μL TE Tween buffer with 2 μL Proteinase K, making sure all tissue is covered.
3. Incubate tubes at 55 °C for 1 h.
4. Incubate tubes at 95 °C for 7 min.
5. Cool tubes to room temperature.
6. Vortex tubes for 1 min to disrupt tissue (*see* **Note 14**).
7. Centrifuge tubes at 1500 × *g* for 5 min to pellet undigested tissue.
8. Dilute the supernatant (a crude DNA preparation) 1 in 10 with sterile water.
9. Perform PCR on the DNA for 35 cycles with the primers detailed in Table 2 and an all-in-one DNA polymerase mix.
10. The resulting product can be resolved visually using a QIAxcel capillary genotyping machine or by agarose gel electrophoresis using standard techniques.

3.3 Phenotypic Examination of Conditional ARKO Mice

To maximize information and minimize time to publication we breed cohorts of mice to standardized ages, normally d0, d21, d35, and d100 (*see* **Note 15**).

3.3.1 Indications of Disrupted Androgen Receptor Action During Necroscopy

Where transgenic mice are produced with a conditional ablation of AR it is possible to get an overview of whether androgen signaling is normal or disrupted during gross dissection of the mouse. Some observations will just be indicative of testicular androgen signaling, others of androgen signaling throughout the whole body.

1. Weigh the whole mouse immediately post-mortem: If decreased body weight is seen in mutants compared to controls, it is suggestive of decreased circulating androgen levels, or a disruption of androgen action in one or more of the organ systems determining body weight. Complete ARKO mice have a "feminized" body weight [50].
2. Measure the anogenital distance (AGD) using an electronic digital caliper: The AGD is the distance from the middle of the anus to the base of the phallus (the penis in males or clitoris in

females). Anogenital distance is a measure of androgen-exposure during the masculinization programming window [84] and is smaller in females and total-ARKO males than it is in normal males.

3. Note whether the testes have descended: If testes are not in the scrotum (as male mice can retract their testes), gentle pressure can be applied above the area of the inguinal canal to determine whether they can be manually moved into the scrotum. Undescended testes may be an indicator of disrupted fetal Leydig cell development as it is induced by the production of INSL3 and testosterone: two fetal Leydig cell products [85].
4. Note the position of the urethral meatus: if it does not emerge through the glans penis but instead on the underside of the penis or the perineum the mouse has hypospadias. Hypospadias is an indicator of disrupted androgen action before birth that has impaired the fusion of the urethral folds during development [84].
5. Note if the mouse has nipple development: retention of nipples in males is indicative of disrupted androgen action [86], as androgens are responsible for suppression of the nipple anlage in males.
6. Dissect the testes from the epididymides and surrounding fat and weigh them: A reduction in testis weight is a strong indication of either a reduction in spermatogenesis or a relative decrease in size of the entire testis if complete spermatogenesis is still present.
7. Check the epididymal morphology: the presence of the highly vascularized initial segment of the epididymis is an indicator of normal lumicrine testosterone secretion from the testis to the epididymis [87]. When a pink colored initial segment is not seen during epididymal dissection this is an indicator that androgen signaling has been disrupted.
8. Dissect and weigh the seminal vesicles (SV): SV weight is plastic and can vary throughout life depending on the circulating testosterone concentrations; as such it has widespread utility as a "biomarker" of circulating androgens. A decrease in SV weight is indicative of a decrease in circulating androgen levels [24].
9. Remove tissues to be used for downstream DNA, protein, or RNA extraction to dry ice after dissection, then freeze at −80°°C as soon as possible. Tissues to be fixed should be removed immediately to Bouin's solution.

3.3.2 Determination of Genomic Ablation of AR in Tissues of Interest by PCR

Once tissues have been collected from conditional ARKO mice, genomic PCR is used to determine if Cre-induced androgen receptor recombination has taken place in individual tissues. Onset of AR ablation can be determined by assaying for genomic recombination at several ages, although a genomic recombination event does not necessarily infer a functional ablation as it may occur before the onset of AR expression in the cell type of interest.

1. Extract genomic DNA from the tissues of interest using a column-based genomic DNA extraction kit according to manufacturer's instructions.
2. Perform PCR on the DNA for 40 cycles with an all-in-one PCR mix and the primers detailed in Table 2.
3. Run the PCR product on either a QIAxcel automated electrophoresis system or by agarose gel electrophoresis.

3.3.3 Determination of Spatial Ablation of AR by Immunohistochemistry

1. When genomic AR ablation in the tissue of interest has been determined, localization of the ablation is pinpointed using immunochemistry following Subheading 3.1.3, but using a rabbit anti-AR antibody as the primary antibody.
2. Perform immunohistochemistry on tissue sections from control littermates are and conditional ARKO mice to determine the localization of normal AR expression in controls and thus ablation in the conditional ARKOs. Ideally processed a control and a knockout in parallel on the same slide.

3.3.4 Histological Analysis of Conditional ARKO Tissues

Once tissue localization of ablation is determined, histological analysis of these tissues is then undertaken to identify any changes in morphology and function due to disruption of androgen receptor signaling. Slide-mounted sections of tissues obtained as in Subheading 3.1.3 are stained using hematoxylin to stain the cell nuclei blue and eosin to stain the cell cytoplasm pink (H&E).

1. Place in a slide rack and move through the following baths (*see* **Note 16**): xylene bath for 5 min to remove paraffin wax, a separate xylene bath for 5 min, 90% ethanol bath for 20 s, and 70% ethanol bath for 20 s.
2. Wash slides under running tap water for 20 s.
3. Place slides in hematoxylin bath for 5 min.
4. Wash slides under running tap water for 20 s.
5. Place slides in acid–alcohol bath for 10 s (to remove nonspecific blue staining).
6. Wash slides under running tap water for 20 s.
7. Place slides in Scott's tap water bath for 20 s (to develop the remaining specific blue color).
8. Wash slides under running tap water for 20 s.
9. Place slides in eosin bath for 5 s, then remove.
10. Immediately wash slides under running tap water for 20 s.

11. Dry slides by blotting rack on absorbent paper in preparation for dehydration (*see* **Note 17**).
12. Place slides in 70% ethanol bath for 20 s.
13. Place slides in 80% ethanol bath for 20 s.
14. Place slides in 95% ethanol bath for 20 s.
15. Place slides in absolute ethanol bath for 20 s.
16. Place slides in a separate absolute ethanol bath for 20 s.
17. Place slides in xylene bath for 5 min (*see* **Note 18**).
18. Place slides in a separate xylene bath for 5 min.
19. Mount slides with Pertex and a coverslip before viewing under a light microscope.

Initial analysis of testicular histology can be based around qualitative observations. Abnormal features to note include lack of one or more spermatogenic cell types to the extreme of a Sertoli cell only phenotype, abnormal appearance of germ cells, lack of a tubular lumen, or Leydig cell hyper/hypoplasia or hyper/hypotrophy [5].

3.3.5 Quantification of Testicular Cell Types

Quantification of the different cell types of the testis is an essential part of determining the effects of a mutant phenotype on normal testicular architecture and function. Although sometimes a qualitative reduction in number of a particular cell type may seem visibly obvious, it could be because of an optical illusion caused by changes in testicular shape and size. Cell counting must be performed for accurate quantification. If it is not possible to determine cell identity by morphology alone, then immunostaining for cell-specific markers can be performed to assist in identification (*see* **Note 19**).

1. Estimate total testis volume using the Cavalieri principle [88].
2. Fix, embed, section, and stain the tissue as in Subheading 3.3.4.
3. Set up a microscope with motorized stage such as an Olympus BX50 microscope and stereological software such as Stereologer.
4. The optical disector technique [89] is used to count the number of Sertoli cells, Leydig cells, spermatogonia, spermatocytes, round spermatids, and elongating spermatids in each testis.
5. Each cell type is identified by previously described criteria [90, 91]. Briefly, intratubular Sertoli cells are recognized by their nuclear shape and tripartite nucleolus. Leydig cells are recognized by their interstitial position, round nucleus, and relatively abundant cytoplasm. Maturation of germ cells occurs from the basal to the luminal surfaces of the seminiferous epithelium so germ cell type can be inferred partly by the position of the cell. Spermatogonia are recognized by their ovoid nucleus and by one flattened surface that rests on the basal

lamina of the tubule. Spermatocytes are recognized by their large round nuclei with darkly staining meiotic chromatin structures and are found closer to the lumen than spermatogonia. Round spermatids are recognized by their smaller rounded nucleus. At later stages, acrosomal development can be seen. Elongating spermatids are the cell type abutting the lumen and are recognized by a small elongated nucleus and the presence of a developing flagellum. Not all spermatogonial cell types will be visible in every tubular cross section.

3.3.6 Analysis of Circulating Hormone Levels

Testosterone production and spermatogenesis are under the control of the pituitary hormones luteinizing hormone (LH) and follicle stimulating hormone (FSH). Levels of these hormones may change in conditional ARKO mice, for example a high LH level combined with low testosterone may indicate compensatory Leydig cell failure or a low level of FSH may indicate impaired spermatogenesis. It is important to note that since the production of these three hormones constitutes the hypothalamic pituitary gonadal (HPG) axis feedback loop, changes in their levels could have a gonadal, hypothalamic, or pituitary cause. Analysis of the location of androgen receptor ablation will help determine the site of the primary defect.

1. Following euthanization, remove blood from mice via cardiac puncture.
2. Coat a 25 G × 5/8 in. gauge needle attached to a 1 mL syringe with heparin by inspiring and expiring a small amount.
3. Insert the needle into the left ventricle of the heart and slowly withdraw blood into the syringe. Between 500 μL and 1 mL can be obtained by this method.
4. Transfer the blood to a 1.5 mL eppendorf tube on ice.
5. Centrifuge the tubes at maximum speed for 10 min to pellet the cellular component.
6. Remove the plasma supernatant and stored at −80 °C until required.
7. Testosterone and other steroid hormones must first be solvent-extracted from plasma before concentration is determined by enzyme-linked immunosorbent assay (ELISA) [92].
8. Place 100 μL of plasma in a glass tube (*see* **Note 20**).
9. In a fume hood, add 2 mL of newly opened diethyl ether to each tube.
10. Cover tubes with a piece of Parafilm.
11. Vortex for 10 min.
12. Centrifuge tubes at 1000 × *g* for 10 min.

13. Place the tubes in a dry ice–ethanol bath for a few minutes to freeze the aqueous bottom phase.
14. Pour off the upper organic phase into a fresh glass tube.
15. Either leave the organic phase to evaporate overnight in a fume hood or use a nitrogen jet machine to dry the samples within a few minutes.
16. Reconstitute the steroids overnight in appropriate ELISA buffer, or buffer appropriate for downstream application (*see* **Note 21**).
17. Testosterone concentration in the extract can then be determined using a commercially available ELISA kit according to the manufacturer's instructions.
18. FSH and LH are glycoprotein hormones that do not need to be pre-extracted from plasma before an ELISA. Concentrations can be determined using commercially available ELISA kits suitable for use with plasma.
19. ELISA plates can be read using a microplate spectrophotometer to determine optical density at a suitable wavelength for the kit.
20. Analyze absorbance data obtained from the spectrophotometer using the freely available ELISA Analysis online application.

3.3.7 Fertility Testing

Disruption of androgen receptor action may result in impaired mating behavior, spermatogenesis, or epididymal function, all of which may decrease the fertility of male mice. Fertility testing is performed on mice as follows:

1. Each male mouse is left in his own cage (*see* **Note 22**) and a single CD1 female mouse added (*see* **Note 23**).
2. Every morning check the female mouse for the presence of a vaginal plug (*see* **Note 24**), indicating that mating had taken place the night before. Record presence or absence of a plug.
3. After plugging, or after 5 days has elapsed (whichever occurs first) remove the female and place another female into the male's cage.
4. Repeat this process until each male had had the opportunity to mate with three females.
5. Remove the females to their own cage, cull at 18 days post plug date and count the number of pups in utero (*see* **Note 25**).
6. Compare conditional ARKOs and control littermates for percentage of females plugged, percentage of pregnancies and number of pups per litter.
7. A lack of plugs and thus pregnancies indicates a defect in mating behavior. When plugs have occurred but not pregnancies, this indicates a likely defect in spermatogenesis or epididymal

function where mature functional spermatozoa are either not being produced or not passing through the epididymis. Analysis of testicular and epididymal histology should assist with determining which of these is the cause.

3.3.8 Assessment of Leydig Cell Maturation Using qRT-PCR

There are two phases of Leydig cell development in rodents. Although both populations are defined by their steroidogenic ability, there is variation in the nature of the steroids secreted and control mechanisms regulating steroidogenesis. Fetal Leydig cells arise during testis development and produce the androgens required to masculinize a male fetus. They stop producing testosterone around the time of birth and their fate in the adult testis is currently unknown. Adult Leydig cell synthesize androgens required for spermatogenesis and adult male reproductive function and arise from a resident peritubular stem Leydig cell population begins around puberty. They arise through several stages of differentiation before becoming mature adult Leydig cells producing high levels of testosterone [93] and each stage expresses different markers and has a different morphology. To assess whether maturation in the Leydig cell lineage has occurred, perform qRT-PCR for the mature adult Leydig cell transcripts *Insl3, Ptgds*, and *Hsd3b6* [94].

1. Isolate RNA from frozen testes using the RNeasy Mini extraction kit with RNase-free DNase on the column digestion kit (Qiagen, Crawley, UK) according to the manufacturer's instructions.
2. Add 5 ng Luciferase mRNA to each testis before RNA extraction as an external standard [95] (*see* **Note 26**).
3. Determine RNA concentration using a spectrophotometer.
4. Prepare random hexamer primed cDNA using the SuperScript VILO cDNA synthesis kit according to manufacturers' instructions. Add an identical amount RNA (up to 2.5 μg) to each reaction.
5. Perform quantitative PCR on triplicates of each cDNA sample using Taqman® Gene Expression Mastermix, and primer/probe sets (*see* Table 4) for both the gene of interest (a Roche Universal Probe Library assay) and luciferase either in separate reactions or in the same reaction as a multiplex (*see* **Note 27**). The end user must optimize reagent concentrations to allow for variation in experimental setup between laboratories.
6. Run reactions on an ABI Prism 7900 machine and analyze using ABI Sequence Detection System software.
7. Use the ΔΔCt method [96] to obtain a relative expression level of the gene of interest compared to a calibrator (*see* **Note 28**).

Table 4
qRT-PCR assays for Leydig cell maturation markers

Assay	Forward primer	Reverse primer	Probe	Label
Hsd3b6	accatccttccacagttctagc	acagtgaccctggagatggt	Roche UPL #95	FAM
Insl3	aagaagccccatcatgacct	tttatttagactttttgggacacagg	Roche UPL #10	FAM
Ptgds	ggctcctggacactacaccta	atagttggcctccaccactg	Roche UPL #89	FAM
Luciferase	gcacatatcgaggtgaacatcac	gccaaccgaacggacattt	tacgcggaatacttc	NED

In conclusion, following this combined approach ensures that the production and validation of the model is convincing and fit for purpose. Empirically validating the Cre Recombinase line to be used is essential, as is demonstrating that neither the Cre Recombinase line nor the floxed line of interest produce a phenotype on their own. Front-loading the breeding part of any project and collecting the oldest samples first minimizes the time to completion, and standardizing the timings of tissue collection across projects permits downstream cross-project comparison of related mutants, with the potential for production of comparison articles or review papers. First-pass analysis of key functional markers, by immunohistochemistry, qRT-PCR, or Western blot can provide a rapid focus to downstream studies. In addition, determination of the cellular composition is essential when examining the testis as any loss of germ cells (which make up greater than 80% of the weight of a testis), will skew any downstream quantitative analysis. Finally, functional analysis of the relevant endocrine environment can also provide key information relevant to downstream study. The guidance, suggestions, and protocols outlined above provide the key starting point for analyses of conditional-ARKO mice, completing them as described provides an excellent framework for further focussed project-specific analyses, and applies equally well to analysis of reproductive tissues from any mouse model generated through conditional gene targeting.

4 Notes

1. Reporter mice can be maintained to any age of interest, but we normally breed to two ages in the first instance; day (d) 2: to assess the sites of expression during fetal testis development, and d80—early adulthood, to determine which cell-types have been targeted during postnatal development.
2. The advantage of this system is that it provides an "at a glance" impression of sites of reporter gene expression throughout the body, including other organs and ectopic sites not normally

examined during detailed histological analysis of your organ of interest (e.g., germ cell expression when working primarily on a liver phenotype) but which may later become extremely relevant. Imaging side-by-side is important in determining presence absence of reporter expression. It is all too easy to ramp up the laser power and convince yourself fluorescence is present when it is not.

3. Bouin's fixative is a mixture of picric acid, acetic acid, and formaldehyde that preserves germ cell nuclear morphology well and allows for easier identification of spermatogenic stages in the testis. Tissue can be fixed for up to 48 h but no shorter than 6 h.
4. Multiple changes of 70% ethanol can remove out excess yellow color from the tissues if required, although we find that this does not interfere with downstream techniques.
5. We use an in-house embedding service to prepare our microtome blocks
6. We have a set of glass baths containing reagents for dewaxing and H&E staining set up in a fume hood. Reagents are changed once a week or more regularly if required.
7. The best pH buffer for epitope retrieval varies for antibodies and tissue samples. Optimization must be performed.
8. This can also be performed in a saucepan-style pressure cooker on an electrical heated plate although no control over temperature settings can be exercised with this method.
9. If a primary antibody raised in goat is to be used, then serum from another animal such as chicken may be used for blocking, then a chicken anti-goat secondary used for detection.
10. The titre we start with for Tyramide optimization is a 1 in 2 dilution series from 1 in 500 to 1 in 8000.
11. Other Tyramide colors are available.
12. Other nuclear counterstains of several different colors are available. We avoid DAPI for use on the testis as it does not stain germ cell nuclei well.
13. Breeding of female conditional ARKO mice requires an extra generation to obtain AR^{flox} homozygotes. In practice, we have found litter size and frequency of birth reduced in F1 ARKO females. F1 female offspring will be mosaic for the conditional ARKO due X-chromosome inactivation. This is an important observation to remember when examining genetically heterozygous female ARKOs—F1 females are part knockout, part wild type.
14. We have found that this step is vital for tissue disruption, it must not be missed out.

15. We begin with the oldest cohort (d100), breeding until we have eight individuals of each genotype per age, as our power calculations based on our previous analysis of similar mutants have demonstrated that this number ensures the detection of a difference in means of 10% at 90% Power in qRT-PCR experiments. By starting with the oldest cohort first, we ensure that any gaps in the datasets can be filled in the shortest possible time, e.g., it is faster to breed mice to d2, than d100. Furthermore, we breed en masse when possible. That is, it costs the same to produce and house the mice whether this is spread over 6 months or 2 years. Maximizing numbers of breeding pairs if possible, ensures the fastest production of primary tissue for analysis. The following protocols are tailored to the phenotypic analysis of conditional ablation of AR in testicular cell types, but the general principles apply both to analysis of conditional ablation of AR in other tissues, and of conditional ablation of other nuclear receptors in testicular cell types. During initial characterization studies, male mice are dissected at the following key ages. Day 0 (to determine whether ablation had occurred embryonically, also a day 0 dissection permits reuse of the mother for additional litters, an e19.5 analysis would require many more breeding females), day 21: early puberty, day 35: first wave of spermatogenesis is complete, day 100: adult Leydig cells have developed and are producing testosterone normally. One year (if deemed relevant for the study): mouse has begun aging.
16. Please *see* **Note 6**.
17. This step is essential to avoid excess water contamination of the dehydrating reagents.
18. If the xylene becomes cloudy on introduction of slides, then there is water contamination in your absolute ethanol baths. Absolute ethanol and xylene baths must be replaced and the slides reintroduced into the first absolute ethanol bath.
19. If the quality of fixation or embedding does not allow for sufficient confidence in identification of each cell type, then immunohistochemistry for cell type markers can be utilized. This would most commonly be required to distinguish Leydig or Sertoli cells from other surrounding cell types. However, it must be emphasized that "markers" may not be expressed in the same spatiotemporal pattern in mutants as they are in controls so caution must be exercised in assessing the correct cell-specific markers to use. A chromogenic label such as 3,3′-diaminobenzidine (DAB) must be used to allow slides to be viewed under a nonfluorescent microscope for counting.
20. Anecdotally, steroids are more likely to be adsorbed by plastic tubes than glass

21. A drop of absolute ethanol may be added to the desiccated steroid to encourage resuspension before addition of buffer, volume of ethanol must not exceed 5 % of the total volume of resuspension buffer.
22. Anecdotally, moving females into the male's established territory enhances mating success.
23. CD1 female mice are chosen for their large litter size.
24. The plug is made of coagulated secretions from the coagulating and vesicular glands of the male. The plug persists inside the vagina for 8–24 h after breeding. The plug is a whitish mass that can be seen by gently lifting the tail of the female and examining her vaginal opening. To make it easier to see you may wish to spread the vulva slightly with a cotton-tipped swab.
25. We wait this long to ensure that pups are viable but without waiting for the mother to give birth.
26. Normalizing to luciferase allows transcripts to be expressed "per testis" which is important when the number of the cell type expressing that transcript differs between control and mutant. In combination with stereological cell counting as detailed in Subheading 3.3.5, this allows changes in transcript level due to changes in cell number and changes in transcript level due to changes in level of expression per cell to be distinguished.
27. The two assays can be distinguished by the different fluorescent wavelengths of the two probes.
28. The ΔΔCt method can only be used if the efficiency of the gene of interest PCR assay is similar to the efficiency of the luciferase PCR assay. If there is a large difference in efficiency, then the standard curve methods must be used.

Acknowledgements

The authors thank members of the Smith group past and present, and our wider collaborators for their efforts in developing and refining the protocols described in this chapter. We also thank Saloni Patel for providing the images used in Fig. 2.
Funding: The work described in this chapter was funded by a MRC Programme Grant Award (G1100354/1) to L.B.S.

References

1. Jorgez CJ, Lin YN, Matzuk MM (2005) Genetic manipulations to study reproduction. Mol Cell Endocrinol 234:127–135
2. Roy A, Matzuk MM (2006) Deconstructing mammalian reproduction: using knockouts to define fertility pathways. Reproduction 131: 207–219
3. Cooke HJ, Saunders PT (2002) Mouse models of male infertility. Nat Rev Genet 3:790–801
4. Jamsai D, O'Bryan MK (2010) Genome-wide ENU mutagenesis for the discovery of novel male fertility regulators. Syst Biol Reprod Med 56:246–259

5. Borg CL, Wolski KM, Gibbs GM, O'Bryan MK (2010) Phenotyping male infertility in the mouse: how to get the most out of a 'non-performer'. Hum Reprod Update 16:205–224
6. O'Bryan MK, de Kretser D (2006) Mouse models for genes involved in impaired spermatogenesis. Int J Androl 29:76–89, Discussion 105–108
7. Yatsenko AN, Iwamori N, Iwamori T, Matzuk MM (2010) The power of mouse genetics to study spermatogenesis. J Androl 31:34–44
8. Sauer B, Henderson N (1988) Site-specific DNA recombination in mammalian cells by the Cre recombinase of bacteriophage P1. Proc Natl Acad Sci U S A 85:5166–5170
9. Sauer B, Henderson N (1989) Cre-stimulated recombination at loxP-containing DNA sequences placed into the mammalian genome. Nucleic Acids Res 17:147–161
10. Hadjantonakis AK, Pirity M, Nagy A (1999) Cre recombinase mediated alterations of the mouse genome using embryonic stem cells. Methods Mol Biol 97:101–122
11. Nagy A (2000) Cre recombinase: the universal reagent for genome tailoring. Genesis 26:99–109
12. Branda CS, Dymecki SM (2004) Talking about a revolution: the impact of site-specific recombinases on genetic analyses in mice. Dev Cell 6:7–28
13. Sauer B (1987) Functional expression of the cre-lox site-specific recombination system in the yeast *Saccharomyces cerevisiae*. Mol Cell Biol 7:2087–2096
14. Nagy A, Mar L, Watts G (2009) Creation and use of a cre recombinase transgenic database. Methods Mol Biol 530:365–378
15. Thyagarajan B, Guimaraes MJ, Groth AC, Calos MP (2000) Mammalian genomes contain active recombinase recognition sites. Gene 244:47–54
16. Semprini S, Troup TJ, Kotelevtseva N, King K, Davis JR et al (2007) Cryptic loxP sites in mammalian genomes: genome-wide distribution and relevance for the efficiency of BAC/PAC recombineering techniques. Nucleic Acids Res 35:1402–1410
17. Kwan KM (2002) Conditional alleles in mice: practical considerations for tissue-specific knockouts. Genesis 32:49–62
18. Gossen M, Bujard H (2002) Studying gene function in eukaryotes by conditional gene inactivation. Annu Rev Genet 36:153–173
19. Lewandoski M (2001) Conditional control of gene expression in the mouse. Nat Rev Genet 2:743–755
20. Smith L (2011) Good planning and serendipity: exploiting the Cre/Lox system in the testis. Reproduction 141:151–161
21. Wang X (2009) Cre transgenic mouse lines. Methods Mol Biol 561:265–273
22. Welsh M, Saunders PT, Atanassova N, Sharpe RM, Smith LB (2009) Androgen action via testicular peritubular myoid cells is essential for male fertility. FASEB J 23:4218–4230
23. Welsh M, Moffat L, Belling K, de Franca LR, Segatelli TM et al (2012) Androgen receptor signalling in peritubular myoid cells is essential for normal differentiation and function of adult Leydig cells. Int J Androl 35:25–40
24. Welsh M, Moffat L, Jack L, McNeilly A, Brownstein D et al (2010) Deletion of androgen receptor in the smooth muscle of the seminal vesicles impairs secretory function and alters its responsiveness to exogenous testosterone and estradiol. Endocrinology 151:3374–3385
25. Welsh M, Moffat L, McNeilly A, Brownstein D, Saunders PT et al (2011) Smooth muscle cell-specific knockout of androgen receptor: a new model for prostatic disease. Endocrinology 152:3541–3551
26. Lobe CG, Koop KE, Kreppner W, Lomeli H, Gertsenstein M et al (1999) Z/AP, a double reporter for cre-mediated recombination. Dev Biol 208:281–292
27. Mao X, Fujiwara Y, Orkin SH (1999) Improved reporter strain for monitoring Cre recombinase-mediated DNA excisions in mice. Proc Natl Acad Sci U S A 96:5037–5042
28. Soriano P (1999) Generalized lacZ expression with the ROSA26 Cre reporter strain. Nat Genet 21:70–71
29. Yamauchi Y, Abe K, Mantani A, Hitoshi Y, Suzuki M et al (1999) A novel transgenic technique that allows specific marking of the neural crest cell lineage in mice. Dev Biol 212: 191–203
30. Tsien JZ, Chen DF, Gerber D, Tom C, Mercer EH et al (1996) Subregion- and cell type-restricted gene knockout in mouse brain. Cell 87:1317–1326
31. O'Gorman S, Dagenais NA, Qian M, Marchuk Y (1997) Protamine-Cre recombinase transgenes efficiently recombine target sequences in the male germ line of mice, but not in embryonic stem cells. Proc Natl Acad Sci U S A 94:14602–14607
32. Sakai K, Miyazaki J (1997) A transgenic mouse line that retains Cre recombinase activity in mature oocytes irrespective of the cre transgene transmission. Biochem Biophys Res Commun 237:318–324

33. De Gasperi R, Rocher AB, Sosa MA, Wearne SL, Perez GM et al (2008) The IRG mouse: a two-color fluorescent reporter for assessing Cre-mediated recombination and imaging complex cellular relationships in situ. Genesis 46: spcone
34. Kawamoto S, Niwa H, Tashiro F, Sano S, Kondoh G et al (2000) A novel reporter mouse strain that expresses enhanced green fluorescent protein upon Cre-mediated recombination. FEBS Lett 470:263–268
35. Luche H, Weber O, Nageswara Rao T, Blum C, Fehling HJ (2007) Faithful activation of an extra-bright red fluorescent protein in "knock-in" Cre-reporter mice ideally suited for lineage tracing studies. Eur J Immunol 37:43–53
36. Mao X, Fujiwara Y, Chapdelaine A, Yang H, Orkin SH (2001) Activation of EGFP expression by Cre-mediated excision in a new ROSA26 reporter mouse strain. Blood 97:324–326
37. Muzumdar MD, Tasic B, Miyamichi K, Li L, Luo L (2007) A global double-fluorescent Cre reporter mouse. Genesis 45:593–605
38. Novak A, Guo C, Yang W, Nagy A, Lobe CG (2000) Z/EG, a double reporter mouse line that expresses enhanced green fluorescent protein upon Cre-mediated excision. Genesis 28:147–155
39. Srinivas S, Watanabe T, Lin CS, William CM, Tanabe Y et al (2001) Cre reporter strains produced by targeted insertion of EYFP and ECFP into the ROSA26 locus. BMC Dev Biol 1:4
40. Vintersten K, Monetti C, Gertsenstein M, Zhang P, Laszlo L et al (2004) Mouse in red: red fluorescent protein expression in mouse ES cells, embryos, and adult animals. Genesis 40:241–246
41. Weber P, Schuler M, Gerard C, Mark M, Metzger D et al (2003) Temporally controlled site-specific mutagenesis in the germ cell lineage of the mouse testis. Biol Reprod 68:553–559
42. Ishikawa TO, Herschman HR (2010) Conditional bicistronic Cre reporter line expressing both firefly luciferase and beta-galactosidase. Mol Imaging Biol 13(2):284–292
43. Lyons SK, Meuwissen R, Krimpenfort P, Berns A (2003) The generation of a conditional reporter that enables bioluminescence imaging of Cre/loxP-dependent tumorigenesis in mice. Cancer Res 63:7042–7046
44. Safran M, Kim WY, Kung AL, Horner JW, DePinho RA et al (2003) Mouse reporter strain for noninvasive bioluminescent imaging of cells that have undergone Cre-mediated recombination. Mol Imaging 2:297–302
45. Lee JY, Ristow M, Lin X, White MF, Magnuson MA et al (2006) RIP-Cre revisited, evidence for impairments of pancreatic beta-cell function. J Biol Chem 281:2649–2653
46. De Gendt K, Swinnen JV, Saunders PT, Schoonjans L, Dewerchin M et al (2004) A Sertoli cell-selective knockout of the androgen receptor causes spermatogenic arrest in meiosis. Proc Natl Acad Sci U S A 101:1327–1332
47. Holdcraft RW, Braun RE (2004) Androgen receptor function is required in Sertoli cells for the terminal differentiation of haploid spermatids. Development 131:459–467
48. Kato M, Shimada K, Saito N, Noda K, Ohta M (1995) Expression of P450 17 alpha-hydroxylase and P450aromatase genes in isolated granulosa, theca interna, and theca externa layers of chicken ovarian follicles during follicular growth. Biol Reprod 52:405–410
49. Notini AJ, Davey RA, McManus JF, Bate KL, Zajac JD (2005) Genomic actions of the androgen receptor are required for normal male sexual differentiation in a mouse model. J Mol Endocrinol 35:547–555
50. Yeh S, Tsai MY, Xu Q, Mu XM, Lardy H et al (2002) Generation and characterization of androgen receptor knockout (ARKO) mice: an in vivo model for the study of androgen functions in selective tissues. Proc Natl Acad Sci U S A 99:13498–13503
51. De Gendt K, Verhoeven G (2012) Tissue- and cell-specific functions of the androgen receptor revealed through conditional knockout models in mice. Mol Cell Endocrinol 352:13–25
52. Walters KA, Simanainen U, Handelsman DJ (2010) Molecular insights into androgen actions in male and female reproductive function from androgen receptor knockout models. Hum Reprod Update 16:543–558
53. Matsumoto T, Shiina H, Kawano H, Sato T, Kato S (2008) Androgen receptor functions in male and female physiology. J Steroid Biochem Mol Biol 109:236–241
54. Verhoeven G, Willems A, Denolet E, Swinnen JV, De Gendt K (2010) Androgens and spermatogenesis: lessons from transgenic mouse models. Philos Trans R Soc Lond B Biol Sci 365:1537–1556
55. Wang RS, Yeh S, Tzeng CR, Chang C (2009) Androgen receptor roles in spermatogenesis and fertility: lessons from testicular cell-specific androgen receptor knockout mice. Endocr Rev 30:119–132
56. Zhou X (2010) Roles of androgen receptor in male and female reproduction: lessons from global and cell-specific androgen receptor knockout (ARKO) mice. J Androl 31:235–243

57. De Gendt K, Verhoeven G (2009) Tissue-selective knockouts of steroid receptors: a novel paradigm in the study of steroid action. In: McEwan IJ (ed) Methods in molecular biology: the nuclear receptor superfamily. Humana Press, New York, pp 237–261
58. Chang C, Chen YT, Yeh SD, Xu Q, Wang RS et al (2004) Infertility with defective spermatogenesis and hypotestosteronemia in male mice lacking the androgen receptor in Sertoli cells. Proc Natl Acad Sci U S A 101:6876–6881
59. Lim P, Robson M, Spaliviero J, McTavish KJ, Jimenez M et al (2009) Sertoli cell androgen receptor DNA binding domain is essential for the completion of spermatogenesis. Endocrinology 150:4755–4765
60. O'Hara L, Smith LB (2012) Androgen receptor signalling in vascular endothelial cells is dispensable for spermatogenesis and male fertility. BMC Res Notes 5:16
61. Tsai MY, Yeh SD, Wang RS, Yeh S, Zhang C et al (2006) Differential effects of spermatogenesis and fertility in mice lacking androgen receptor in individual testis cells. Proc Natl Acad Sci U S A 103:18975–18980
62. Xu Q, Lin HY, Yeh SD, Yu IC, Wang RS et al (2007) Infertility with defective spermatogenesis and steroidogenesis in male mice lacking androgen receptor in Leydig cells. Endocrine 32:96–106
63. Simanainen U, McNamara K, Davey RA, Zajac JD, Handelsman DJ (2008) Severe subfertility in mice with androgen receptor inactivation in sex accessory organs but not in testis. Endocrinology 149:3330–3338
64. O'Hara L, Welsh M, Saunders PT, Smith LB (2011) Androgen receptor expression in the caput epididymal epithelium is essential for development of the initial segment and epididymal spermatozoa transit. Endocrinology 152:718–729
65. Krutskikh A, De Gendt K, Sharp V, Verhoeven G, Poutanen M et al (2011) Targeted inactivation of the androgen receptor gene in murine proximal epididymis causes epithelial hypotrophy and obstructive azoospermia. Endocrinology 152:689–696
66. Kaftanovskaya EM, Huang Z, Barbara AM, De Gendt K, Verhoeven G et al (2012) Cryptorchidism in mice with an androgen receptor ablation in gubernaculum testis. Mol Endocrinol 26:598–607
67. McInnes KJ, Smith LB, Hunger NI, Saunders PT, Andrew R et al (2012) Deletion of the androgen receptor in adipose tissue in male mice elevates retinol binding protein 4 and reveals independent effects on visceral fat mass and on glucose homeostasis. Diabetes 61:1072–1081
68. Yu IC, Lin HY, Liu NC, Wang RS, Sparks JD et al (2008) Hyperleptinemia without obesity in male mice lacking androgen receptor in adipose tissue. Endocrinology 149:2361–2368
69. Altuwaijri S, Chuang KH, Lai KP, Lai JJ, Lin HY et al (2009) Susceptibility to autoimmunity and B cell resistance to apoptosis in mice lacking androgen receptor in B cells. Mol Endocrinol 23:444–453
70. Lai KP, Lai JJ, Chang P, Altuwaijri S, Hsu JW et al (2013) Targeting thymic epithelia AR enhances T-cell reconstitution and bone marrow transplant grafting efficacy. Mol Endocrinol 27:25–37
71. Lai JJ, Lai KP, Chuang KH, Chang P, Yu IC et al (2009) Monocyte/macrophage androgen receptor suppresses cutaneous wound healing in mice by enhancing local TNF-alpha expression. J Clin Invest 119:3739–3751
72. Lin HY, Xu Q, Yeh S, Wang RS, Sparks JD et al (2005) Insulin and leptin resistance with hyperleptinemia in mice lacking androgen receptor. Diabetes 54:1717–1725
73. Yu IC, Lin HY, Liu NC, Sparks JD, Yeh S et al (2013) Neuronal androgen receptor regulates insulin sensitivity via suppression of hypothalamic NF-kappaB-mediated PTP1B expression. Diabetes 62:411–423
74. Raskin K, de Gendt K, Duittoz A, Liere P, Verhoeven G et al (2009) Conditional inactivation of androgen receptor gene in the nervous system: effects on male behavioral and neuroendocrine responses. J Neurosci 29:4461–4470
75. Notini AJ, McManus JF, Moore A, Bouxsein M, Jimenez M et al (2007) Osteoblast deletion of exon 3 of the androgen receptor gene results in trabecular bone loss in adult male mice. J Bone Miner Res 22:347–356
76. Chiang C, Chiu M, Moore AJ, Anderson PH, Ghasem-Zadeh A et al (2009) Mineralization and bone resorption are regulated by the androgen receptor in male mice. J Bone Miner Res 24:621–631
77. Sinnesael M, Claessens F, Laurent M, Dubois V, Boonen S et al (2012) Androgen receptor (AR) in osteocytes is important for the maintenance of male skeletal integrity: evidence from targeted AR disruption in mouse osteocytes. J Bone Miner Res 27:2535–2543
78. Dubois V, Laurent MR, Sinnesael M, Cielen N, Helsen C et al (2014) A satellite cell-specific knockout of the androgen receptor reveals myostatin as a direct androgen target in skeletal muscle. FASEB J 28(7):2979–2994
79. Ophoff J, Van Proeyen K, Callewaert F, De Gendt K, De Bock K et al (2009) Androgen signaling in myocytes contributes to the maintenance of muscle mass and fiber type regulation

but not to muscle strength or fatigue. Endocrinology 150:3558–3566
80. Chambon C, Duteil D, Vignaud A, Ferry A, Messaddeq N et al (2010) Myocytic androgen receptor controls the strength but not the mass of limb muscles. Proc Natl Acad Sci U S A 107:14327–14332
81. Welsh M, Sharpe RM, Walker M, Smith LB, Saunders PT (2009) New insights into the role of androgens in wolffian duct stabilization in male and female rodents. Endocrinology 150:2472–2480
82. Kawano H, Sato T, Yamada T, Matsumoto T, Sekine K et al (2003) Suppressive function of androgen receptor in bone resorption. Proc Natl Acad Sci U S A 100:9416–9421
83. Willems A, De Gendt K, Deboel L, Swinnen JV, Verhoeven G (2011) The development of an inducible androgen receptor knockout model in mouse to study the postmeiotic effects of androgens on germ cell development. Spermatogenesis 1:341–353
84. Welsh M, Saunders PT, Fisken M, Scott HM, Hutchison GR et al (2008) Identification in rats of a programming window for reproductive tract masculinization, disruption of which leads to hypospadias and cryptorchidism. J Clin Invest 118:1479–1490
85. Hughes IA, Acerini CL (2008) Factors controlling testis descent. Eur J Endocrinol 159(Suppl 1):S75–S82
86. Imperato-McGinley J, Binienda Z, Gedney J, Vaughan ED Jr (1986) Nipple differentiation in fetal male rats treated with an inhibitor of the enzyme 5 alpha-reductase: definition of a selective role for dihydrotestosterone. Endocrinology 118:132–137
87. Abe K, Takano H (1989) Cytological response of the principal cells in the initial segment of the epididymal duct to efferent duct cutting in mice. Arch Histol Cytol 52:321–326
88. Mayhew TM (1992) A review of recent advances in stereology for quantifying neural structure. J Neurocytol 21:313–328
89. West MJ, Slomianka L, Gundersen HJ (1991) Unbiased stereological estimation of the total number of neurons in the subdivisions of the rat hippocampus using the optical fractionator. Anat Rec 231:482–497
90. O'Shaughnessy PJ, Monteiro A, Abel M (2012) Testicular development in mice lacking receptors for follicle stimulating hormone and androgen. PLoS One 7:e35136
91. Russell LD (1990) Histological and histopathological evaluation of the testis, vol xiv. Cache River Press, Clearwater, Fl, 286 p
92. Handelsman DJ, Jimenez M, Singh GK, Spaliviero J, Desai R et al (2015) Measurement of testosterone by immunoassays and mass spectrometry in mouse serum, testicular, and ovarian extracts. Endocrinology 1:400–405
93. Chen H, Ge RS, Zirkin BR (2009) Leydig cells: from stem cells to aging. Mol Cell Endocrinol 306:9–16
94. O'Shaughnessy PJ, Johnston H, Willerton L, Baker PJ (2002) Failure of normal adult Leydig cell development in androgen-receptor-deficient mice. J Cell Sci 115:3491–3496
95. Baker PJ, O'Shaughnessy PJ (2001) Expression of prostaglandin D synthetase during development in the mouse testis. Reproduction 122:553–559
96. Livak KJ, Schmittgen TD (2001) Analysis of relative gene expression data using real-time quantitative PCR and the 2(-Delta Delta C(T)) Method. Methods 25:402–408

Index

A

B

C

D

E

F

G

H

I

L

Iain J. McEwan (ed.), *The Nuclear Receptor Superfamily: Methods and Protocols*, Methods in Molecular Biology, vol. 1443, DOI 10.1007/978-1-4939-3724-0,

Printed in the United States
By Bookmasters